Nanotechnology Science and Technology Series

MAGNETIC PROPERTIES AND APPLICATIONS OF FERROMAGNETIC MICROWIRES WITH AMORPHOUS AND NANOCRYSTALLINE STRUCTURE

NANOTECHNOLOGY SCIENCE AND TECHNOLOGY SERIES

National Nanotechnology Initiative: Assessment and Recommendations
Jerrod W. Kleike (Editor)
2009. ISBN 978-1-60692-727-4

Nanotechnology Research Collection - 2009/2010. DVD edition
James N. Ling (Editor)
2009. ISBN 978-1-60741-293-9

Nanotechnology Research Collection - 2009/2010. PDF edition
James N. Ling (Editor)
2009. ISBN 978-1-60741-292-2

Safe Nanotechnology in the Workplace
Nathan I. Bialor (Editor)
2009. ISBN 978-1-60692-679-6

Strategic Plan for NIOSH Nanotechnology Research and Guidance
Martin W. Lang (Author)
2009. ISBN: 978-1-60692-678-9

Nanotechnology in the USA: Developments, Policies and Issues
Carl H. Jennings (Editor)
2009. ISBN: 978-1-60692-800-4

New Nanotechnology Developments
Armando Barrañón (Editor)
2009. ISBN: 978-1-60741-028-7

Electrospun Nanofibers and Nanotubes Research Advances
A. K. Haghi (Editor)
2009. ISBN: 978-1-60741-220-5

Nanostructured Materials for Electrochemical Biosensors
Umasankar Yogeswaran, S. Ashok Kuma and Shen-Ming Chen
2009. ISBN: 978-1-60741-706-4

Magnetic Properties and Applications of Ferromagnetic Microwires with Amorpheous and Nanocrystalline Structure
Arcady Zhukov and Valentina Zhukova
2009. ISBN 978-1-60741-770-5

Nanotechnology Science and Technology Series

Magnetic Properties and Applications of Ferromagnetic Microwires with Amorphous and Nanocrystalline Structure

ARCADY ZHUKOV
AND VALENTINA ZHUKOVA

Nova Science Publishers, Inc.
New York

For permission to use material from this book please contact us:
Telephone 631-231-7269; Fax 631-231-8175
Web Site: http://www.novapublishers.com

LIBRARY OF CONGRESS CATALOGING-IN-PUBLICATION DATA

Zhukov, Arcady, 1963-
Magnetic properties and applications of ferromagnetic microwires with amorpheous and nanocrystalline structure / Arcady Zhukov and Valentina Zhukova.
p. cm.
Includes index.
ISBN 978-1-60741-770-5 (hardcover)
1. Amorphous substances--Magnetic properties. 2. Ferromagnetic materials. 3. Nanocrystals. I. Zhukova, Valentina, 1962- II. Title.
TA418.9.A58Z48 2009
620.1'1297--dc22

2009017366

Published by Nova Science Publishers, Inc. ✢ New York

CONTENTS

PREFACE

Recent advances in technology involving magnetic materials require development of novel advanced magnetic materials with improved magnetic and magneto-transport properties. Additionally the tendency on miniaturization of the modern magnetic sensors and devices stimulates development of such magnetic materials with reduced dimensionality. Certain progress has been recently achieved in fabrication of novel magnetic nano-materials (thin films, nanowires, nano-dots…), but at the same time quite sophisticated technology should be used but in many occasions the magnetic properties of these materials are rather poorer than such properties of bulk magnetic materials (amorphous ribbons, wires, sintered materials…) and the fabrication process is much more expansive and complex. On the other hand certain industrial sectors, like magnetic sensors, microelectronics, security etc, need cheap materials with reduced dimensionality but still with high magnetic properties (particularly enhanced magnetic softness). Therefore magnetic materials with outstanding magnetic characteristics and reduced dimensionality recently gained much attention.The aim of this book is to present the most relevant aspects concerning the fabrication process following by sections devoted to present significant results on the remagnetization process, compositional dependence and processing of glass-coated mirocwires with amorphous, nanocrystalline and granular character in order to design their magnetic properties.

ACKNOWLEDGMENTS

Authors wish to acknowledge support by EU ERA-NET programme under project DEVMAGMIWIRTEC (MANUNET-2007-Basque-3) by the Basque Government under Saiotek 08 METAMAT project. We wish also to acknowledge contribution of our collaborators from Magnetic Group of Basque Country University, from Inst. Phys., Fac. Sci., UPJS, Kosice, Slovakia, from Moscow State University from Plymouth Univeristy, from Insitute of Material Science of Madrid, from Oviedo University, from AmoTec and MFTI (M. Ipatov, M. Ilyn, A. Chizhik, J. M. Blanco, C. Garcia, J.J. del Val, J. Gonzalez, R. Varga, V. Samsonova, L. Panina, D. Mahnovskyi, M. Vazquez, C.Luna, B. Hernando, V. Prida, K. Garcia, A. Torcunov, V. Larin et al)

Chapter 1

INTRODUCTION

Recent advances in technology involving magnetic materials require development of novel advanced magnetic materials with improved magnetic and magneto-transport properties. Additionally the tendency on miniaturization of the modern magnetic sensors and devices stimulates development of such magnetic materials with reduced dimensionality. Certain progress has been recently achieved in fabrication of novel magnetic nano-materials (thin films, nanowires, nano-dots…), but at the same time quite sophisticated technology should be used but in many occasions the magnetic properties of these materials are rather poorer than such properties of bulk magnetic materials (amorphous ribbons, wires, sintered materials…) and the fabrication process is much more expansive and complex [1,2]. On the other hand certain industrial sectors, like magnetic sensors, microelectronics, security etc, need cheap materials with reduced dimensionality but still with high magnetic properties (particularly enhanced magnetic softness). Therefore magnetic materials with outstanding magnetic characteristics and reduced dimensionality recently gained much attention.

Regarding their magnetic properties magnetic materials are usually classified into two main groups with either hard or soft magnetic properties. Soft magnetic materials can be magnetized by relatively low magnetic fields, and when the applied field is removed, they return to a state of relatively low remanent magnetism. Soft magnetic materials typically exhibit coercivities, H_c, below 400 A/m (5 Oe). Soft magnetic behavior is important in any application involving a change in magnetic induction [1,2].

Hard magnetic materials retain a large amount of residual magnetism after exposure to a magnetic field. These materials typically have coercivities, Hc, above 10 kA/m (125 Oe). These materials are used principally as a source of a magnetic field.

Among these families of modern magnetic materials a family of thin wire with reduced geometrical dimensions (of order of 1-30 μm in diameter) gained importance within last few years [1-4]. These thin wires combine excellent soft magnetic properties (with coercivities till 4 A/m) with attractive magneto-transport properties (Giant Magneto-impedance effect, GMI, Giant Magneto-resistance effect, GMR) and unusual re-magnetization process in positive magnetostriction compositions exhibiting magnetically bistable behaviour and quite fast domain wall propagation [2-6].

Regarding existing and future applications of such thin wires, a number of magnetic sensors based on giant magneto-impedance (GMI) effect and stress-impedance (SI) effect with the C-MOS IC circuitry and advantageous features comparing with conventional magnetic sensors have been reported [7,8]. Main possible applications reported in Ref. [7,8]

are related with the detection of the magnetic fields, small weights and vibrations, and such branches of the industry as the car industry and medicine are main consumers of these sensors. Among the almost 100 applications, the following sensors are currently in progress: portable digital display of the terrestrial magnetic field, brain tumor sensor, secondary current sensor for the induction motor control, car passing measurement and recording disk, finger-tip blood vessel pulsation, mechanoencephalogram sensor, etc [7,8].

On the other hand new types of stress- tuneable and temperature-tuneable composite materials based on thin ferromagnetic wires with the effective microwave permittivity depending on an external dc magnetic field, applied stress or temperature recently have been introduced [9, 10]. Such composites consist of short pieces of conductive ferromagnetic wires embedded into a dielectric matrix. The short wire inclusions play a role of "the elementary scatterers", when the electromagnetic wave irradiates the composite and induces a longitudinal current distribution and electrical dipole moment in each inclusion. These induced dipole moments form the dipole response, which can be characterized by some complex effective permittivity. The later may have a resonance or relaxation dispersion caused by the strong current distribution along a wire, which depends on the wire high frequency surface impedance. In the vicinity of the resonance frequency any variations in the wire surface impedance result in a large change of the current distribution, and hence in the dipole moment of each inclusion and the effective permittivity on the whole. For a ferromagnetic conductive wire, the surface impedance may depend not only on its conductivity but also on the dc external magnetic field and tension through the so-called magneto-impedance effect (MI). Therefore, the dispersion of the effective permittivity can be tuned from a resonance type to a relaxation type, when a sufficient magnetic field or tensile stress is applied to the composite sample. A number of applications have been proposed, including the stress-sensitive media for remote non-destructive health monitoring of different structures, temperature –dependent media and selective microwave coatings with field-dependent reflection/transmission coefficients [9,10]. The important advantage of such applications is that the soldering problems are avoided because of the wireless detection of the signals. It is worth mentioning that thin wires with stress sensitive magnetic anisotropy exhibiting stress sensitive GMI effect and SI effect are quite necessary for designing of such composites.

On the other hand magnetic domain wall propagation becomes a hot topic of research because of its use in magnetic devices (such as Magnetic Random Access Memory, Integrated Circuits, Hard Disks, etc.) to transmit the information along the magnetic wire of submicrometer diameter [11,12].

In the case of magnetic microwires with positive magnetostriction constant the magnetic bistability characterized by the appearance of rectangular hysteresis loop at low applied magnetic field has been observed [13-15]. This magnetic bistable behavior is related to the presence of a single Large Barkhausen Jump, which was interpreted as the magnetization reversal in a large single domain [13]. Such peculiar remagnetization process is quite interesting for various applications [16-17]. It is important, that singla and large Barkhausen jump mis observed above some critical fields Regarding the critical length, it can be correlated well with the demagnetizing factor [13] indicating that the closure domains penetrate from the wire ends inside the internal axially magnetized core destroying the single domain structure. For the case of commercially available amorphous wires (with diameter about 120 μm) this critical length is about 7 cm, which is quite inconvenient for use in magnetic micro-sensors and microelectronics.In glass-coated microwires with diameter about 10 μm this critical length is much shorter (about 2 mm) which is quite suitable for applications in microsensors. This rectangular hysteresis loop also disappears when the magnetic field is below some critical value denominated as the switching field [13]. Such rectangular hysteresis loop was interpreted in terms of nucleation or depinning of the reversed domains inside the internal single domain and the consequent domain wall propagation [14]. Perfectly rectangular shape of the hysteresis loop has been related with a very high velocity of such domain wall propagation. It is demonstrated by few methods that the remagnetization process of such magnetic microwire starts from the sample ends as a consequence of the depinning of the domain walls and subsequent DW propagation from the closure domains [14-16].

Quite surprising results such as exiting of the domain wall propagation below the switching field [14], negative critical propagation field [15] and supersonic domain wall propagation [16] have been reported. It is worth mentioning that best soft magnetic properties have been achieved in amorphous and/or nanocrystalline thin wires. Crystallization usually results in degradation of magnetic properties related with appearance of numerous pinning centres (grain boundaries, magnetocrystalline anisotropy…)[17,18]. Although in some cases the character of the hysteresis loop keeps similar to that of the amorphous state [18].

It is now nearly 50 years since the first metallic glass (amorphous metal) was produced by rapid quenching from the liquid state by I.S. Miroshnitchenko and I.V. Salli [19] and later by P. Duwez et al [20]. This proved to be starting point in opening up new fields of research in material science, magnetism and technology. Novel materials with unique combinations of properties (magnetic, mechanical, corrosion etc) such as metastable crystalline phases and structures, extended solid solubilities of solutes with improved mechanical and physical properties, nanocrystalline, nanocomposite and amorphous materials have been introduced [19,20]. Technological development of the fabrication technique and studies of the structure, glass formation ability and thermodynamics and magnetism of amorphous alloys were intensively performed in 60-th-70-th. Main attention in 60-th-70-th was paid to amorphous ribbons, as extensively described in few review papers and books [21-23] .

Most scientific, commercial and technological interest has been paid to magnetically soft amorphous and nanocrystalline magnetic materials. Initially it was not clear if the ferromagnetic ordering could exist in amorphous solids because of lack of atomic ordering. In 1960 Gubanov theoretically predicted in 1960 that amorphous solids would be ferromagnetic [24]. Later a number of experimental works show that the amorphous alloys obtained by rapid-quenching of the melt based on 3d-transition metals exhibit excellent soft magnetic materials with very low coerticity value, high magnetic permeability and relatively high saturation magnetization [23]. Such magnetic softness has been related with the absence of magnetocrystalline anisotropy in these alloys [23]. Particularly, combination of excellent soft magnetic properties of amorphous ribbons obtained by the melt-spinning technique with high wear and corrosion resistance made them very attractive for development of novel soft magnetic materials and for development of the applications in magnetic sensors, magnetic recording head and microtransformer industries [25]. Concerning the magnetic ordering in materials with structural disorder (particularly amorphous material) there are some fundamental questions related to the existence of such magnetic order. In fact, considering the ferromagnetic interactions of the magnetic materials it can be immediately considered a ferromagnetic structure. But the origin of magnetic anisotropy in structurally disordered materials is unclear. Magnetic moments tend to arrange their orientations parallel to each other via exchange interactions; this they do when lying along a magnetic easy axis which is in the same direction at every point in the material. However, if the easy axis orientation fluctuates from site to site, a conflict between ferromagnetic coupling and anisotropy arises. As long as we imagine lattice periodicity, a ferromagnetic structure is a consequence of ferromagnetic exchange interactions, the strength of the anisotropy being irrelevant. In this situation we are assuming a major simplification, namely: the direction of the easy axis is uniform throughout the sample. With this simple picture we present crucial questions related to the influence of an amorphous structure on magnetic order.

Regarding the magnetic ordering in amorphous and nanocrystalline materials, it is assumed by most of researchers that it determines mostly by two anisotropies contributions: exchange and local anisotropy. The exchange contribution arises from the electron-electron correlations. The mechanism of the electrostatic interactions between electrons has no relation to structural order and is sensitive only to overlapping of the electron wave functions. With respect to magnetic anisotropy is also originated by the interaction of the local electrical field with spin orientation, through the spin-orbit coupling. Therefore, magnetic anisotropy is also a local concept. Nevertheless, the structural configuration of magnetic solids exerts an important influence on the macroscopic manifestation of the local anisotropy. As a consequence, when the local axes fluctuate in orientation owing to the structural fluctuation (amorphous and nanocrystalline materials as examples), calculations of the resultant macroscopic anisotropy becomes quite difficult. In the case of amorphous magnets the usual approach when a magnetic order connected to a lattice periodicity is not applicable. The structure of amorphous materials should be considered as similar to the liquid, because the crystallization process can not develop because of high quenching rate of the liquid. Therefore these magnetic materials can be defined as solids in which the orientation of local symmetry axes fluctuate with a typical correlation length l = 1nm. The local structure can be characterized by a few local configurations with icosahedral, octahedral, and trigonal symmetry. These structure units have randomly distributed orientation. The local magnetic anisotropy would be larger in these units with lower symmetry. It is remarkable that in these types of structures the correlation length, l, of such fluctuation is typically corresponds to the correlation length of the structure and ranges from 1 nm (amorphous) to 10 nm (nanocrystals) and 1 mm (polycrystals).

In fact, in order to obtain amorphous structure, an ingot of molten alloy must undercool below the equilibrium crystallization temperature before crystallization begin because of the existence of an energy barrier to the formation of nuclei [27]. Eventually, when the cooling rate is sufficiently high, crystallization is suppressed because of insufficient time for significant growth or nucleation. In practice additions of the components which decrease the crystallization temperature is quite important. Therefore usually amorphous alloys have a phase diagram with deep eutectic (for example alloys transition metal-metalloid).

Fluctuations in the interatomic distances associated with the amorphous structure should also contribute to some degree of randomness in the magnetic interactions of the magnetic moments. Nevertheless, such randomness is expected not to affect the magnetic behavior qualitatively [23,28]. Moreover, random distribution of the orientation of the easy axis drastically affects to the magnetic properties. Random anisotropy model developed by R. Alben et al. [29] provides a successfully explanation that how the correlation length, l, exerts a relevant influence on magnetic structure. Usually amorphous magnetic alloys exhibit estremely soft magnetic behaviour because of absence of magnetocrystalline anisotopy, grain boundaries, and crystalline structure defects.

As mentioned above, although crystallization usually results in degradation of magnetic softness of amorphous alloys, in some cases crystallization can improve magnetically soft behavior. This is the case of so-called "nanocrysalline" alloys obtained by suitable annealing of amorphous metals. These materials have been introduced in 1988 by Yoshizawa et al. [30] and later have been intensively studied by a number of research gropus [31-36]. Researh and technological interest in such nanocrystalline alloys, denominated also as "Finemet" (in the case of Fe-rich nanocrystalloine alloys) arose from extremely soft magnetic properties combined with high saturation magnetization. This nanocrystalline structure of partially crystalline amorphous precursos is observed in Fe-Si-B with small additions of Cu and Nb. It is widely assumed that the role of these small additions of Cu and Nb results in inhbiting of the grains nucleation and decreasingn of the grain growth rate [31-36]. Such soft magnetic character is though to be originated because the magnetocrystalline anisotropy vanishes and the very small magnetostriction value when the grain size approaches 10 nm [31-34]. As was theoretically estimated by Herzer [32,37] , average anisotropy for randomly oriented α-Fe(Si) grains is negligibly small when grain diameter does not exceed about 10 nm. Thus, the resulting magnetic behavior can be well described with random the anisotropy model [31,33-35,37]. According to this model, low coercivity in the nanocrystalline state is ascribed to small effective magnetic anisotropy (K_{eff} around 10 J/m^3). It is worth mentioning that lower magnetic anisotropy of nanocrystalline state (around 300 – 500 J/m^3) as-compared with the amorphous precursors (~1000 J/m^3) down to that obtained in the nanocrystalline alloys is not sufficient to explain such low values of coercivities typical for the nanocrystalline state [36-38]. The enhancement of the magnetic softness should be ascribed to the decrease of the microstructure-magnetization interactions. These interactions originating in large number of coupled magnetic moments suggest a relevant role of the magnetostatic interactions as well in the formation of these coupled units [36,37]. In addition to the suppressed magnetocrystalline anisotropy, low magnetostriction values provide the basis for the superior soft magnetic properties observed in particular compositions. Low values of the saturation magnetostriction are essential to avoid magnetoelastic anisotropies arising from internal or external mechanical stresses. The increase of initial permeability with the formation of the nanocrystalline state is closely related to a simultaneous decrease of the saturation magnetostriction. Partial crystallization of amorphous alloys leads to an extention of the frequency range where the permeability presents high values [39]. These high values of magnetic permeability in extended frequency range are quite useful in many technological applications involving the use of ac magnetic fields.

It is remarkable that a number of researchers have investigated the effect of the substitution of Fe in the $Fe_{73.5}Cu_1Nb_3Si_{13.5}B_9$ alloy composition (so-called *Finemet)* by an additional alloying element, like Co or Ni, in order to improve the magnetic properties [40-46]. Thus, it was shown that substitution of Fe by Al in the classical *Finemet* composition led to a significant decrease in the minimum of coercivity, $H_{cmin} \approx 0.5$ A/m, achieved after partial devitrification [44], although the effective magnetic anisotropy field was quite large (around 550 A/m) [46]. Quite soft magnetic behaviour and GMI effect were observed in Finemet nanocrysalline ribbons where Fe has bee partially subsituted by Co [43].

Although amorphous Fe-, Co- and Ni-based ribbons are slightly more expansive comparing with conventional soft magnetic materials, such as sendast, ferrites and supermalloys (mostly due to the significant content of Co and Ni), they found considerable applications in transformers (400 Hz), ac powder distributors (50 Hz), magnetic recording as a magnetic heads and mostly in magnetic sensors. The main reason for using amorphous alloys as soft magnetic materials is a saving of the electric energy wasted by magnetic cores. Besides, the combination of high magnetic permeability and good mechanical properties of amorphous alloys may be successfully used in magnetic shielding and in magnetic heads [47]. Thus, production about 3 millions heads per year in Japan in mid-1980s has been reported [47].

Starting from 90-th a novel family of amorphous materials – amorphous wires have been introduced [48-50]. First generation of amorphous wire deals with typical diametr around 125 μm in diameter, obtained by the so-called in-rotating-water quenching technique. This kind of materials ehxibit a number of unusual magnetic properties. Thus, the magnetostrictive compositions exhibit rectangular hysteresis loop, while best magnetic softness is observed for the nearly-zero magnetostriction composition. Their main technological interest is related to the magnetic softness in nearly-zero magnetostriction composition, magnetic bistability in non-zero magnetostriction compositions and GMI effect [50-52]. This GMI effect consists in the large change of the electric impedance of a magnetic conductor when is subjected to an axial dc magnetic field. It has been recognized that the large sensitivity of the total impedance of a soft magnetic conductor at low magnetic fields and high frequencies of the driven ac current originates from the dependence of the transverse magnetic permeability upon the dc magnetic field and the skin effect. Besides, the magnetoelectric effects (inverse Wiedemann and Matteucci effects) have been widely used too.

These magnetic properties of amorphous wires attraced great interest in 90-th mostly to great posibilities for use in sensor applications (mostly owing to magnetic bistability and GMI effect). Besides, the mechanism of the remagnetization process changes with the sign of magnetostriction constant, being such mechanism as of nucleation of domain walls for Fe-rich composition and as of propagation for Co-rich wires.

Last advances in magnetic materials are based on the miniaturization of modern magnetic materials (see comparison in Figure 1). The alternative technology of rapid quenching – Taylor Ulitovski method to produce thinner metallic wires (in the order of 1 to 30 μm in diameter) covered by an insulating glass coating has been widely employed for fabrication of ferromagnetic microwires coated by glass (see photo in Figure 2) since middle of 90-th [2-4, 53-55]. It should be mentioned that the technology itselt for fabrication of metallic glass-coated microwires is known since 50-th and first reports on fabrication and properties of glass-coated microwires have been published in 70-th (see more details in Chapters 2 and 4). The great advantage of these mirowires is that the obtained diameter could be significantly lower than in the case of amorphous wires produced by in-rotating water method. But thier magnetic properties are also quite different from "thicker"amorphous wires. Thus although like in the case of "thicker wires" it was observed that Fe-rich compositions with positive magnetostriction constant show generally rectangular hysteresis loop, Co-rich negative magnetostrictive compositions have almost non-hysteretic magnetization curves and glass coating removal results in appearance of magnetic bistability [53, 55]. This is because the glass coating introduces additional internal stresses due to the difference between the thermal

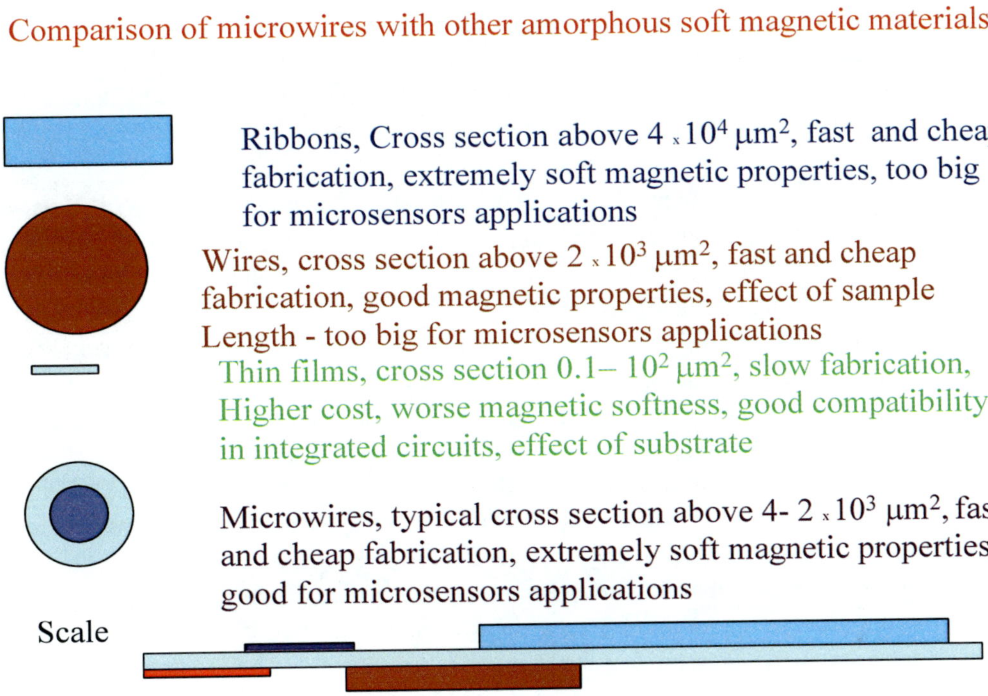

Figure 1. Scheme illustrating comparison of microwires with other soft magnetic materials.

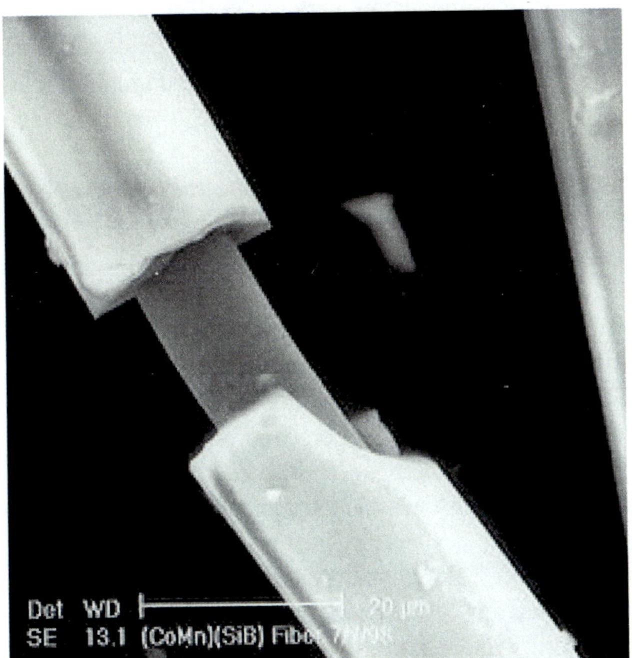

Reprinted with permission from [2] A. Zhukov, J. González, M. Vázquez, V. Larin and A. Torcunov, Nanocrystalline and Amorphous Magnetic Microwires, Encyclopedia of Nanoscience and Nanotechnology, American Scientific Publishers Vol. 6 (2004) p.365-387.

Figure 2. Micrograph of the glass-coated microwire.

expansion coefficients of glass coating and metallic nucleus. Therefore, microwires of the same composition can show different magnetic properties because of the different magnetoelastic energy. Depending on the thickness of the glass coating, the switching field (applied magnetic field necessary to observe magnetic bistability) is generally one order of magnitude higher than for melt-spun wires.

On the other hand, when the magnetostriction constant, λ_S, is close to zero, a great variety of magnetic effects can be observed, depending on the sign of λ_S. It has been found that the hysteresis loop changes from unhysteretic for slightly negative magnetostriction constant to rectangular for positive magnetostriction [56].

The internal stresses results to be as the main source of magnetic anisotropy in amorphous and nanocrystalline materials due to the magnetoelastic coupling between magnetization and internal stresses through magnetostriction. I most of amorphous materials the main origin of these internal stresses is related with high quenching rate and solidification process. Thus solidification usually proceeds from the surface. In the case of amorhpous ribbons additionally the cooling rate is higher from the contact side of ribbon, i.e. where the ribbons contacts with the quenching drum [2, 23, 27]. As a result, the morphology of contact and free surfaces of ribbons are quite different too. In the case of glass-coated microwires, the glass coating introduces an additional magnetoelastic contribution to the magnetic anisotropy acting as a new parameter determining the magnetization processs [2-4]. The origin of these additional stresses is determinded by significant difference of the thermal expansion coefficients of the glass and the metal [2-5, 13, 17]. In the case of low positive magnetostriction constant some of experimental results were expplained assuming the change of the sign of the magnetostriction constant under the effect of internal stresses [2].

Consequently, these materials result to be very interesting for field and stress-sensing elements because the Fe-rich amorphous alloys exhibiting high magnetostriction values (λ_s = 10^{-5}) and therefore many of magnetic parameters (i.e.: magnetic susceptibility, coercive field,...) are extremely sensitive to the applied stresses.

The other source of magnetic anisotropy can be the shape anisotropy. In the case of amorphous ribbons and wires this shape anisotropy can be significant. Thus, for Fe-rich amorphous wires with diameter of about 120 μm significant effect of the samples length on hysteresis loop has been observed for the samples with 7 cm long [13]. In this sense, thin wires produced by Taylor- Ulitovsky technique have clear advantage: demagnetizing factors does not affect their magnetic behaviour if the sample length exceeds few mm [13].

Certianly all aforementioned method has advantages and disadvantages. From point of view of magnetic properties amorphous ribbons and wires produced by melt-quenching have usually superior soft magnetic properties. The process itself is cheap. But aforementioned shape anisotropy influence makes them almost insuitable for micro-sensors applications. On the other hand usually magnetic softness of thin films produced by sputtering, laser ablation or deposition is rather poorer [2] and technology itself is more sofisticated. Taking into account these considerations, thin microwires with reduced demensionality (comparable with that of thin films, see comparison in Figure 1) can play an important role for the case of microsensor applications. In this case additional efforts are needed in order to avoid additional magnetoelastic anisotropy coming from the glass. Recent advances in tailoring of magnetic properties of glass-coated microwires demonstrate, that excellent magnetic sofness can be achieved even for quite thin microwires (with diameter of the order of 10 μm or even less) selecting well the chemical composition and/or applying themal treatment [4, 57].

Recent studies demonstrated that Taylor- Ulitovsky technique for fabrication of glass-coated thin wires allows obtaining semi-hard magnetic behavior by adequate thermal treatment of amorphous precursos [58] and thin microwires with granular structure exhibiting gian magnetoresistance (GMR) effect [59].

It should be mentioned, that certain afferts has been made in order to achieve magnetically soft thin wires by different methods: by cold drawn method and using alternative fabrication of thin wires without glass (melt extraction method) [60-64]. Both these method allow fabrication of thinner microwires (till 20 μm by cold-drawn and till 40 μm – by melt extraction [63,64]). Some interesting results on enhanced magnetic softness have been obtained, but still the diameter cannot be reduced below 20 μm.

Chapter 2

FABRICATION METHOD

The fabrication method denominated in most of modern publications as a modified Taylor-Ulitovsky and/or quenching-and-drawing method [2-4, 55, 60-66] is actually well-known since 60-th. Initially this method has been used mostly for fabrication of non-magnetic microwires (for example Cu), but the casting method is essentially the same. This method is based essentially on direct casting from the melt, as schematically depicted in Figure 3.

The method essentially consists of simultaneous drawing of the composite microwire (metallic nucleus inside the glass capillary) through the quenching liquid (water or oil) jet onto rotating bobbins.

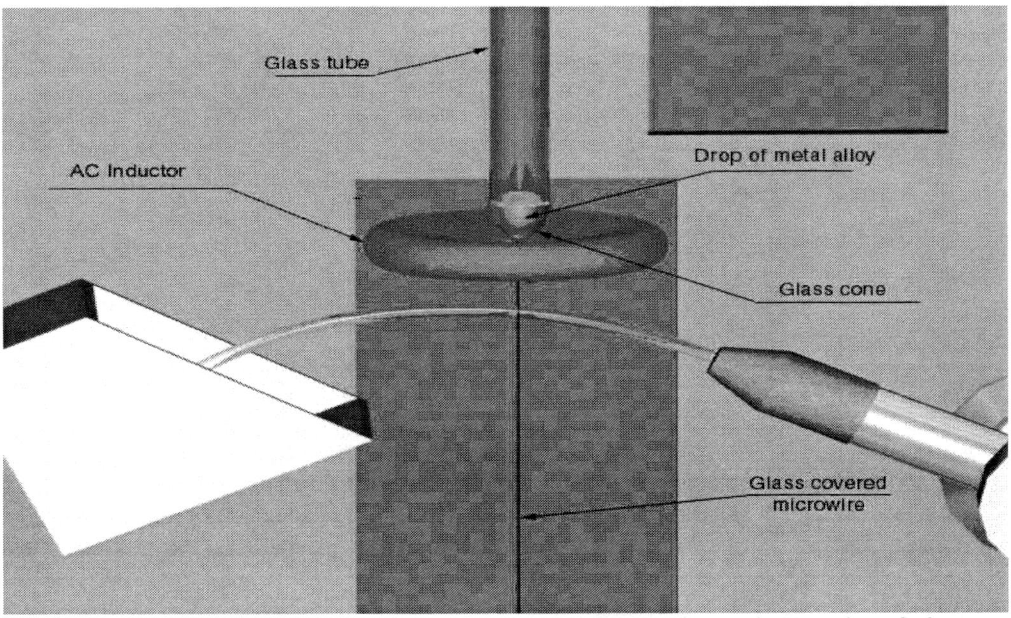

Reprinted with permission from [70], V. S. Larin et al., "Preparation and properties of glass-coated microwires", J. Magn. Magn. Mater. 249(1-2) pp.39-45 (2002). © 2002, Elsevier Science (figure 1).

Figure 3. Schematic drawing of microwire fabrication process by the Taylor–Ulitovsky method.

In the laboratory process, an ingot containing few grams of the master alloy with the desired composition is put into a Pyrex-like glass tube and placed within a high frequency inductor heater. The alloy is heated up to its melting point, forming a droplet. While the metal melts, the portion of the glass tube adjacent to the melting metal softens, enveloping the metal droplet. A glass capillary is then drawn from the softened glass portion and wound on a rotating coil. At suitable drawing conditions, the molten metal fills the glass capillary and a microwire is thus formed where the metal core is completely coated by a glass shell. The amount of glass used in the process is balanced by the continuous feeding of the glass tube through the inductor zone, whereas the formation of the metallic core is restricted by the initial quantity of the master alloy droplet. The microstructure of a microwire (and hence, its properties) depends mainly on the cooling rate, which can be controlled by a cooling mechanism when the metal-filled capillary enters into a stream of cooling liquid (water or oil) on its way to the receiving coil.

The main advantages of this method of microwire fabrication are [60]:

i. repeatability of microwire properties at mass-production;
ii. extended range of variation in parameters (geometrical and physical);
iii. fabrication of continuous long pieces of microwire up to 10 000 m;
iv. control and adjustment of geometrical parameters (inner core diameter and glass thickness) during the fabrication process

In spite of many advantages this method meets some complexities related with peculiarities of metallurgical processes in rapidly quenching of the composite material and effect of electromagnetic field of inductor on alloy ingot and stability of the process. All these peculiarities affect magnetic and structural properties of cast microwires.

Some of these peculiarities are briefly revised below.

2.1. Chemical and Metallurgical Processes Related with Interaction of the Ingot Alloy and the Glass

There are various problems related with influence of the high-frequency magnetic field of the inductor and high temperature on quality and chemical homogeneity of cast microwire. These problems are relatively poorly described in modern publications devoted mostly to studies of magnetic properties of microwires. But some of these problems are closely related with the wire properties reproducibility in the case of mass production and/or homogeneity of these properties along the microwire length and magnetic properties itself.

Thus it is important to take into account the interaction between the alloy and the glass coating. The character of this interaction depends on chemical composition of the ingot as well as on type of glass used for the casting [66].

Particularly, small particles of the glass can be involved by the molten metal ingot moving under the effect of the high-frequency magnetic field of the inductor. Therefore the chemical composition of the ingot at the beginning of the casting process can be different from the alloy composition from that at the end of casting process. Such glass inclusions appearing in the molten ingot can penetrate within the molten alloy inside the capillary during the casting and even interrupt the casting process [66]. It should be mentioned that interaction of glass with molten alloy depends on chemical composition of this alloy. Therefore, since the technology has been developed mostly for casting of non-magnetic alloys, this phenomena has been studied in details for non-magnetic alloys. Thus, for Cu appearance of CuO and Cu_2O, SiO, Si_2O and $CuSiO_3$ has been detected both in droplet of molten Cu and inside the cast microwires. The content of SiO and Si_2O in the Cu droplet increases with time. The content of CuO and Cu_2O and $CuSiO_3$ inside the cast microwire is quite inhomogeneous and depends on the weight of the molten droplet. At the same time, much more homogeneous distribution of oxides has been observed during casting of Ag microwire.

The other hand the interfacial layer between the metallic nucleus and glass coating has been observed and studied for the case of non-magnetic alloys. This interfacial layer has been observed by various methods, among them by the phase analysis using chemical etching for the removal of the glass coating by HF [66]. Correlation between the content of non-metallic inclusions in the interfacial layer and ρ–ratio between metallic nucleus diameter, d, and total diameter, D ($\rho = d/D$) has been observed: the content of non-metallic inclusions decreases increasing the ρ–ratio [66].

Several assumptions have been discussed regarding the origin of the interfacial layer. Thus, among the origins of the interfacial layer the following considerations have been discussed:

1) The connection between the glass and the metallic nucleus through the formation of the series of solid solutions. This assumption is confirmed by the results of elements distribution within the microwire performed using X-ray microanalysis. Considerable change of the Fe, Co, Ni and Si along the microwire radius within few μm has been found between the metallic nucleus and glass coating. The thickness of the interfacial layer in this case can achieve few μm.
2) Uncompensated molecular forces on the interface between the glass and the metallic nucleus. In this case the thickness of the interfacial layer should be less than 0,1 μm
3) Formation of stable chemical compounds with the structure different (from crystallographic point of view) from that of the interacting materials. Appearance of these chemical compounds is possible in the case if the metallic alloy and glass have an electron affinity and if the outer (valent) electronic shells are unfilled. The thickness of the interfacial layer is of the same order (few μm) as in the case of formation of solid state solutions, but the interface layer is more visible, like it was observed by microscopy in few cases [66].

Taking into account aforementioned regarding the interfacial layer, glass-coated microwire consist of metallic nucleus, glass insulating shell and interfacial layer which should be in an ideal case as more homogeneous as possible. It is considered that better homogeneity is achieved if this interfacial layer consists of solid state solution forming from oxides of elements of the glass coating and metallic nucleus. For the case of Cu it is considered that this interfacial layer in most appropriate case should consist on $CuSiO_3$ [66]. It is worth mentioning that the character of the interfacial layer as well as the properties of microwires depends on properties and character of interaction between selected alloy and selected glass as well as on geometry (metallic nucleus diameter, d, and glass coating thickness). Such behaviour relatively well studied for microwire with metallic nucleus from Cu, Ag, but poorly studied in microwires with metallic nucleus from ferromagnetic material.

The other source of instability of properties of cast microwire is related with gas content inside the microwire. The sources of the gas are: the atmosphere, the gas impurities in the alloy and the glass. Some content of oxygen and/or hydrogen (about 5 cm^3/100 g each other) and even nitrogen has been detected. Gaseous precipitations can cause the metallic nucleus deformation, cracks and even discontinuities. Chemical reactions of hydrogen with the oxides of the metals can result in appearance of water bubbles inside the metallic nucleus.

Generally the content of the gases can be reduced or at least stabilized using protective atmosphere (Ar) or vacuum.

2.2. Electromagnetic and Electro-Hydrodynamic Phenomena in the System Of Inductor- Ingot

Main peculiarity of the Taylor-Ulitovsky method is melting of the ingot placed inside the glass tube by the HF currents from the inductor with consequent filing of the capillary forming form the softened glass and filled by the molten alloy. This electro-dynamic interaction determining the shape of the molten ingot and temperature regimes significantly affects the casting process and should be also taking into account. Theoretical aspects of this phenomenon are quite sophisticated and related with the shape of the used inductors. But in simplified case describing the behaviour of the sphere under the effect of single and/or few loops with electrical current have been analyzed [66]. Thus at some conditions the HF irradiation of the inductor interacts with the induced irradiation of the molten ingot and oscillations of the molten ingot can take place. The characteristics of these oscillations depend on the weight of the ingot and therefore during the fabrication process the oscillation can appear or disappear. From the theoretical point of view this system has been analyzed in quasi-static approximation, describing the behaviour of the spherical ingot with electrical conductivity, σ_e=0, under effect of the current loop in media with permittivity and magnetic permeability, $\varepsilon=\mu=1$ [66]. This approach allowed estimating the release of power within the spherical ingot.

It is worth mentioning that the most critical point during the casting of microwires is the bottom part of the molten ingot. At special conditions determined by the shape of the HF inductor, electromagnetic field inside the molten ingot, weight of the ingot, viscosity etc, the equilibrium shape of the ingot can be achieved. At these conditions the casting process is homogeneous.

Achievement of these conditions depend on many parameters, such as releasing power to the ingot, shape of the inductor, frequency of the electromagnetic irradiation created by the inductor, conductivity of the alloy and the weight of the metal.

Determination of these conditions allows establishing reliable casting process.

The casting process is carried out at certain temperature at which the aloys is mold and the surrounding glass tube is soft.

The peculiarity of the system is that the metallic ingot is surrounded by the viscous glass shell and therefore the system can be described as the vibrating system with damping. This results in considerable energy bleeding from the molten ingot and hinders the extraction of the molten ingot from the inductor [66]. This behaviour in fact depends on the ingot mass.

Therefore, in according to the theoretical approach [66] there is a critical weight for the ingot. This critical weight depends on the chemical composition of the ingot.

Analyzing the conditions of formation of cast microwire one can subdivide the forces acting during the process on quasi-static and dynamic. Among quasi-static forces the weight of the glass and metallic alloy, electromagnetic pressure, capillary pressure should be mentioned. Among the dynamic forces the tensile stresses arising from drawing of the microwire on bobbin, forces related with inertia of moving glass and metallic alloy and related with solidification of both metal and glass should be related. In some cases, some of these contributions can be measured. Thus, tensile force, F_{σ}, arising from the drawing has been measured as a function of the velocity of casting, V_c, for the case of Cu microwire. Thus, varies from 350 till 700 Dn when V_c increase from 350 till 700 m/min.

From the point of view of modern micro- sensors applications [7] it is important to produce quite thin microwires with good magnetic softness. Various attempts have been performed to estimate the limit of the method from the point of view of the theory of the surface phenomena, neglecting the wetting between the glass and the metallic alloy. The final result depends on the coefficients of the surface tension of metal, glass and interfacial layer, thickness of interfacial layer, melting temperature and external pressure. [66]

It should be mentioned again, that these aspects have been well studied for the case of non-magnetic alloys and metals, but for the novel family of magnetic materials based on Fe, Co and Ni these problems should be addressed.

2.3. Thermal Conditions of Formation of Cast Microwire

The point of view of properties of cast microwire the thermal conditions play important role. It should be mentioned that there are 2 main problems related with thermal conditions: thermal phenomena inside the molten ingot under effect of electromagnetic field of the inductor and thermal conditions related with solidification of cast microwire.

Main efforts related with the first point have been performed on determination of the temperature distribution inside the ingot and the effect of the metallic bar dimensions on the temperature distribution inside the metallic ingot. It was found the case of Cu, that increasing the diameter of the bar from 2 till 6 mm the temperature inside the ingot decreases on 10-35% [66].

The conditions of the cooling of cast microwire determine the crystalline structure of produced microwire.

2.4. Parameters of the Casting Process and Their Limits

2.4.1. Casting Rate

There are several factors that restrict the upper limit of the casting rate. This rate is limited, first, by the ability of metal to follow the moving glass. Its actual limit can be calculated by assuming that it is necessary to accelerate the initially still metal and glass up to the operating rates [8,9]. A force F must then be imposed to the metal to fill the capillary.

Assuming a conical shape at the end of the glass capillary, as shown in Figure 3, the casting rate V_c (which corresponds to the average speed of the accelerated metal <*dz*/*dt*>) is found from the condition that the work done by this force equals the kinetic energy of metal filling the capillary:

$$(v\rho)V_c^2/2 = Fz \tag{1}$$

where $v = \pi z(R^2 + r^2 + Rr)/3$ is the volume of the conical portion (geometrical parameters are given there), and ρ is the density of metal. The value of r is determined by the final size of the metal core. The limit of the casting rate V_c^* can be obtained from (1), where F coincides with the adhesion force F_a for metal/glass interface:

$$V_c^* = \sqrt{\frac{6F_a}{\pi\rho\,(R^2 + r^2 + Rr)}} \tag{2}$$

Thus, the limiting rate of casting grows with an increase in the adhesion for a given metal/glass pair, and drops with an increase in the core radius.

The limiting casting rate associated with the dynamical viscosity of the glass (which is a function of temperature) and the acceleration of moving glass can be found where the total force F is represented as

$$F = F_\eta(T) + F_a \tag{3}$$

$$F_\eta(T) = k(T)\eta(T)V_c S_{gl}/h \tag{4}$$

where $F_\eta(T)$ is the tensile force acting on the microwire, which is related to the dynamical adhesion of glass at casting temperature, $k(T)$ is the coefficient of resistance to fluidity, $\eta(T)$ is the dynamical viscosity, S_{gl} is the cross-section of the glass coating, h is the length of the stretching path, and F_a is the force related to the acceleration of the glass portion. Thus, taking Eqs. (3) and (4) the limiting rate V_c^* is:

$$V_c^* = 1/\left(k(T)\,\eta\,(T)S_{gl} + F_a\right)^{1/2} \tag{5}$$

2.4.2. Diameter of a Microwire

Typical limits for the metallic core diameter are between 0.8 and 30 μm, while the thickness of the coating is in the range of 2 and 15 μm. It should be noted, that even during the stationary casting process, there is some variation (although not great) in diameter of the metallic nucleus and in the glass coating thickness along the wire length. The experimental evaluation of this variation in terms of the diameter of metallic nucleus and the total diameter as a function of the sample length is shown in Figure 4 for the case of Fe-Si-B / Pyrex microwire.

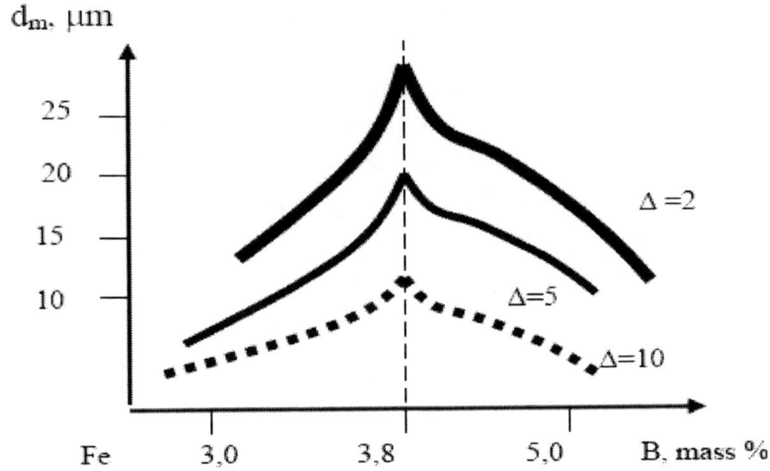

Reprinted with permission from [70], V. S. Larin et al. "Preparation and properties of glass-coated microwires", J. Magn. Magn. Mater. 249(1-2) pp.39-45(2002). © 2002, Elsevier Science (figure 5).

Figure 4. Maximum value of the diameters of the metallic nucleus to obtain microwire (FeSi system) with amorphous microstructure for a range of glass-coating thickness.

The geometrical characteristics of the microwire depend on the physical properties of the glass and metal, the diameter of the initial glass tube, and the parameters of the heating inductor. It has been proven, that the strongest effect on the geometry of a microwire comes from the glass properties. Therefore, the radius of the metal core is estimated to be [67-70]:

$$r_m = 0{,}9\left(\frac{\eta^5}{\sigma^2\upsilon_\lambda}\right)^{\frac{1}{3}}\frac{1}{d}\frac{\eta^5}{\sigma^2\upsilon_\lambda}f(\upsilon,\eta,\sigma,d) \tag{6}$$

with η as

$$\eta = 6\sqrt{6\rho}\left(\frac{\omega^2}{N_0}\right)^{\frac{1}{3}}M^{-5/6}d^{4/3}T^{1/2}\left(\cosh\frac{2\lambda_i^0}{\gamma RT}-1\right) \tag{7}$$

where R is the universal gas constant, ω is the internal volume of molecules in liquid per gram-molecule, N_0 is the Avogadro constant, λ^0_i is the internal hidden evaporation heat per gram-molecule, γ is the co-ordination number, M is the molecular weight, σ is the surface tension, d is the density, $f(\upsilon, \eta, \sigma, d)$ is a factor which can be experimentally evaluated (it depends weakly on the casting rate), and

$$\upsilon_\lambda = \frac{2\nu r_2^2}{\sqrt{3}\rho_2^3} \pm \sqrt{\left(\frac{2\nu r_2^2}{\sqrt{3}\rho_2^3}\right)^2 - \frac{2\sigma}{r_2 d}} \qquad (8)$$

where $\nu = \eta/d$ is the kinematical density. Taking Eqs. (1) to (8), it can be concluded that there exists a limiting casting rate for fabricating a microwire with the smallest possible diameter.

The limit for the upper diameter limit can be obtained at low casting rates, since the contribution of $\upsilon_\lambda^2/2$ is neglected, yielding

$$r_{m\max} = \left(\frac{\sqrt{3}\sigma}{2\nu d}\right)^{1/3} \frac{\rho_2}{\upsilon_\lambda^{1/3}} \qquad (9)$$

Thus, the diameter of a microwire produced by Taylor-Ulitovsky method has both upper and lower limits depending on the rate of casting.

2.4.3. Composition of the Metal Core

The present method allows pure metals such as Ag, Au, Cu, Ni, Sn, Pb, Pt to be used. Semiconductors such as Ge, Si, or Bi can also be cast. The alloys of these metals can be also processed, but not over the whole range of compositions. For example, the content of Cr in Ni and Co based alloys can be up to 20%, the content of Mo, W, V in Ni, Co, and Fe based alloys can be up to 8 - 10 at. %. Some limitations relate to the melting temperature, which must not be too high. On the other hand, rare earth metals and metals which react with the glass and atmosphere (Ti, Cr, Mo, W, Nb, Al, Na, La, Nd) are not suitable for this technology.

Of principle interest is the fabrication of magnetic microwires with soft magnetic characteristics for which Fe, Ni and Co are the alloy's main elements. In the case of amorphous microwire, elements such as Si, B, C, and Al are added to enable the amorphicity of the alloys.

Metastable supersaturated solid solutions can be obtained for immiscible metal systems in a solid and even liquid state, such as Cu-Co, Cu-Co–Ni, Ag-Fe, Ag-Co. As-cast microwire can exhibit a structure of a supersaturated solid solution of transition metals in Cu or Ag. After suitable thermal treatments small single domain particles of Co or Fe can precipitate.

Recently giant magnetoresistance (GMR) effect attritubed to the scattering of the electrons on grain boundaries betweeen ferromagnetic grains and paramagnetic matrix has been obserevd in such compositions [5,71,72].

Mixted crystalline-amorphous structure can be also obtained by thermal annealing of inicially amorphous microwires [73,74] as well as in-prepared state using specially designed mixed compositions like (FeSiB)-Cu, (CoFeSiB)-Cu. These compositions exhibit unusual magnetic properties with irregular hysteresis loops. Additionally, recently novel magnetically soft compositions with small additions of Cr of based on Ni with ferromagnetic-paramagnetic phase transitions near room temperature have been introduced [75,76]

2.5. Microstructure of Cast Microwires

The properties of microwires are related to a great extent to the microstructure of the metal core. The analysed method of casting offers the possibility of rapid cooling and solidification of liquid metal in a non-equilibrium process, allowing the formation and existence of metastable metallic phases. The microstructure of rapidly cooled metal depends on its chemical composition, the melting temperature of metal, and the cooling rate. Depending on the critical cooling rates, different metastable phases can be formed, ie supersaturated solid solution, microcrystalline and amorphous phases. In certain cases, a mixed microstructure consisting of nano or micro-crystals embedded in the amorphous matrix can be formed as well.

2.5.1. Cooling Rate of the Metal Core

The cooling rate V_t depends both on the geometry of the microwire (i.e., diameter d of the metal core and thickness t of the glass cover), and on the parameters of the technological process related to thermal conductivity [67].

$$V_t = \alpha(T_2 - T_k)\,(d + t)/[(\rho_m d^2 C_m + \rho_g C_g (d\,t + t^2)] \qquad (10)$$

where ρ_m , ρ_g, C_m, C_g are the density and specific heat of metal and glass, respectively, T_k is the temperature of the cooling liquid, α is thermoconductivity parameter at the boundary microwire/cooling mediium, and T_2 is the temperature of metal when entering the cooling zone. The value of T_2 can be deduced from

$$T_2 = \alpha^* T_1 \exp[1/V1\,(T_1 - T_a)]\,/\,h \qquad (11)$$

where T_1 is the temperature at the base of the droplet, T_a is the air temperature, V is the casting rate, and α^* is a heat transmission parameter at the movement of a cylinder in the gaseous medium. The characteristic cooling rates of the metal core in the gaseous medium ranges from 10^2 to10^4 K^{-1}, whereas in a stream of liquid ranges between 10^5 and 10^7 K^{-1}.

2.5.2. Metastable Phases in Cast Microwire

Various types of metallic alloys' metastable systems can be prepared by increasing the cooling rate (now, so-called quenching rate). In principle, metal-metalloid, metal-metal eutectic type, and alloys with a limited solidification range in a solid state are significant.

The "critical quenching rate", V_{cr} is as follows:

$$V^*_{cr.} = (T_l - T_g)/t \qquad (13)$$

where T_l and T_g are the liquids and devitrification temperatures respectively, and t is the maximum time at which any part of the material quenches from T_l to T_{g}. A "critical quenching rate" can be defined as the minimum quenching rate necessary to obtain any metastable structure at a given thermal exchange coefficient. The critical quenching rate can be characterised by the maximum metallic nucleus diameter (at a given glass-coating thickness and thermal exchange coefficient, α) at which any of the metastable phases can be obtained.

As an example, Figure 4 illustrates the maximum diameters of the metallic nucleus necessary to obtain an amorphous microstructure in Fe-B microwires. Microwires with diameters below the corresponding curves are obtained in an amorphous state.

Chapter 3

MECHANICAL PROPERTIES

Most early reports on glass coated microwires were devoted to the mechanical properties [77-80]. Thus, effect of annealing and crystallization on tensile strength and Young modulus was studied in amorphous Fe-rich and stainless steel microwires produced by the Tailor-Ulitovski technique [78,79]. High tensile strength (around 250 kg/mm^2) was observed in amorphous $Fe_{80}P_{16}C_3B_1$ microwires as-compared with the crystalline microwires (around 130 kg/mm^2). Such tensile strength is found to decrease increasing the metallic nucleus diameter. This tendency has been attributed to the higher cooling rate with decreasing of the metallic nucleus diameter. On the other hand, the Young modulus of amorphous samples (8000 kg/mm^2) is found to be much lower than of crystalline (13000 kg/mm^2) samples of the same composition. Fracture surfaces of the amorphous microwires consist of a smooth region produced by shearing and a vein pattern [79]. It was observed later, that additions of Si, Co or Cr results in increasing of the averaged tensile strength [80].

Recently interest in studies of mechanical properties has been renewed mostly owing to the promising technological applications of glass coated microwires. In this way, tensile strength and fracture character has been recently studied in CoMnSiB microwires [81]. A wide number of samples of the CoMnSiB series have been studied and the average load for fracture has been found to range between 20±3 g, with a maximum of 30 g and a minimum of 15 g. The tensile strength is then 624±94 MPa, with maximum and minimum values of 936 and 468 MPa, respectively. The tensile strength in amorphous state seems to be due to the metallic nucleus of the microwire. The glass coating does not seem to contribute significantly in resistance to fracture. The glass around the point of fracture breaks, probably because of the sound wave produced by the rupture or the metallic nucleus, which is then visible.

Moreover, samples with higher tensile strength seem to break in a slightly different mode than the ones breaking under lower stress. Thus sometimes the fracture plane is basically perpendicular to the axis of the microwire. On the other hand, another group of samples broke under stresses over the average, show a more complex situation. These samples show some structure in the surface of the metallic nucleus of the microwire that could have improved the adhesion of the metallic nucleus to the coating and increased strength slightly.

An interesting correlation of mechanical and magnetic properties has been observed in "*finemet*"- type microwires during the devitrification [82]. As mentioned above, annealing of initially amorphous $Fe_{73.4}Cu_1Nb_{3.1}Si_{13.4}B_{9.1}$ microwires results in the two-step crystallization process. First a fine nanocrystals α–Fe (Si) of grain size around 10 nm have been precipitate at temperatures around 550-600 °C. At temperature above 600 °C, the beginning of the precipitation of iron borides (with grain size larger than 50 nm) has been found [82].

Such as described elsewhere [2,18] (more detaled description will be given in in Sec. *IV. 2*) typically a decrease of coercivity has been observed at annealing temperatures below 650 °C with two minimums of coercivity, H_c, at about 400-450 °C and at 550-650 °C. Such soft magnetic behaviour has been ascribed to the first crystallization process leading to the precipitation of fine nanocrystals α–Fe (Si) of grain size around 10 nm with 70-80% of relative volume, such it has been widely reported for FINEMET ribbons (see for example ref. [30-32]). An abrupt increase of the coercivity is shown by the samples treated at temperature above 600 °C. The beginning of such increase should be connected to the precipitation of iron borides (with grain size larger than 50 nm) which implies a magnetic hardening character (Figure 5).

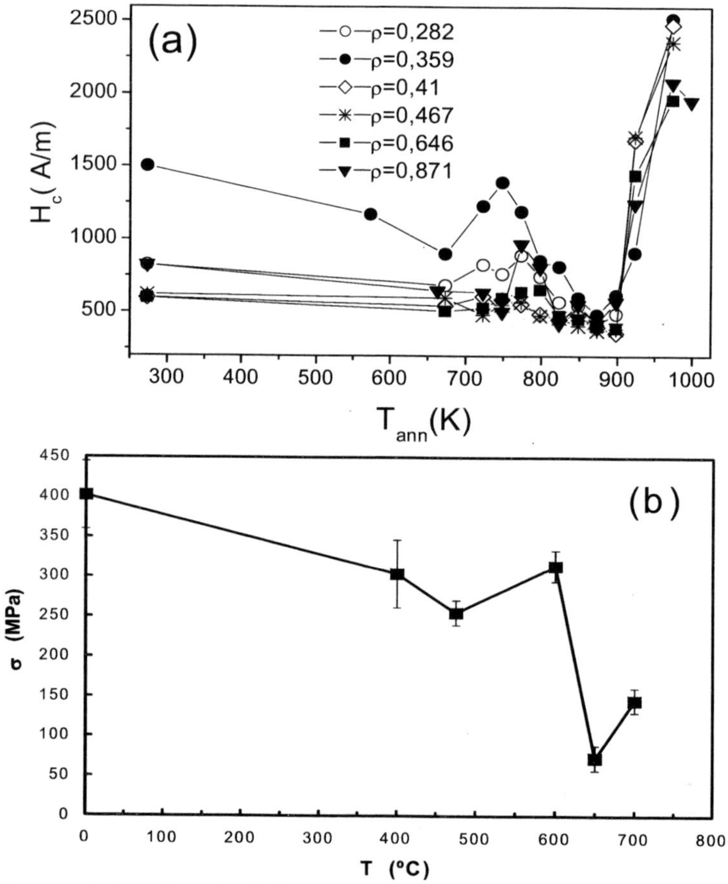

(a)Reprinted with permission from Ref. [136] A. Zhukov, "Glass-coated magnetic microwires for technical applications" J. Magn. Magn. Mater, 242–245 P1, 216-223, 2002, Copyright Elsevier B.V. (2002) (fig 4a); (b) Reprinted with permission from Ref. [82] V. Zhukova et al., "Correlation between magnetic and mechanical properties of devitrified glass-coated $Fe_{71.8}Cu_1Nb_{3.1}Si_{15}B_{9.1}$ microwires" J. Magn. Magn. Mater. 249(1-2), 79-84, 2002, Copyright Elsevier B.V (2002) (fig 4).

Figure 5. (a) Dependence of coercivity on the annealing temperature, T_{ann}, for $Fe_{71.8}Cu_1Nb_{3.1}Si_{15}B_{9.1}$ with the ratio *d*/*D* ranging from 0.282 to 0.646, (b) Dependence of tensile yield on the annealing temperature in $Fe_{71.8}Cu_1Nb_{3.1}Si_{15}B_{9.1}$ microwire.

Tensile yield, σ_y, has been measured for samples annealed at different temperatures. These results are presented in Figure 5. Maximum σ_y was found for the as fabricated sample (402 MPa). Then, certain correlation can be outlined with respect to the dependence of coercivity on annealing temperature. Like in the case of coercivity, tensile yield, σ_y, decreases with T_{ann} with a small local minimum around T_{ann} = 450 °C. Some increase of σ_y has been observed at T_{ann}=575 °C. Finally strong deterioration of the tensile yield has been observed at T_{ann}>650 °C and small increase at T_{ann} =700 °C. The samples annealed at the two highest temperatures were extremely brittle. The σ_y results in figure 5b for these types of microwires correspond only to the maximum measurable value. In most opportunities, samples would fracture even under very careful handling.

The character of the sample fracture has been also studied for the samples annealed at few annealing temperatures by means of SEM microscopy. The annealing does not deteriorate significantly the glass coating as well as metallic nucleus. The fracture character is complex. The metallic nucleus seems to be responsible for the tensile yield, with possible contribution of plasticity.

It is possible that the thermal treatment deteriorates the glass-coating interface due to the difference in thermal expansion coefficients), having a slight effect on the tensile yield.

For the annealing temperatures higher than 600 °C, the character of the fracture is different. If the metal core is now brittle, the glass coating can be responsible for the sample fracture and the sample breaks immediately after the coating itself.

The correlation of the mechanical properties with magnetic behaviour is to be related with structural changes introduced by the annealing. The precipitation of the second crystalline phase leads to a completely crystalline structure, introducing strong internal stresses and change in the character of the chemical bonding. This crystallization process makes the sample brittle [9-12].

Chapter 4

AMORPHOUS MICROWIRES AND THEIR MAGNETIC PROPERTIES

INTRODUCTION
SOFT MAGNETIC MATERIALS

Soft ferromagnetic materials are those magnetic materials with high permeability, low coercivity and low hysteresis loss [1], which can be used to amplify the flux density generated by a magnetic field. The available range of magnetic properties of soft magnetic materials is continually being expanded. This amounts to reduction in coercivity, increase in permeability and consequently a decrease in hysteresis loss. Normally this can be accomplished through reduction in anisotropy and reduction in domain wall pinning. The latter can be strongly affected by the impurity content.

Magnetic Properties

Permeability

Permeability is the most important parameter for soft magnetic materials since it indicates how much magnetic induction *B* is generated by the material in a given magnetic field strength *H*.

Initial permeabilities of all magnetic materials range from $\mu \approx 10^6$ [1] in materials such as amorphous alloys down to as low as $\mu r \approx 1$ in some of the permanent magnets. It is known that initial permeability and coercivity have in broad terms a reciprocal relationship, so that materials with high coercivity necessarily have a low initial permeability and vice versa.

Coercivity

Coercivity is the parameter which is used to distinguish hard and soft magnetic materials. Traditionally a material with a coercivity of less than 1000 A /m is considered magnetically 'soft'. A material with a coercivity of greater than 10,000 A/m is considered magnetically 'hard'. Low coercivities are achieved in nickel alloys such as permalloy in which the coercivity can be as low as 0.4 A/m [1,2]. In amorphous and nanocrystalline materials coercivity also achieves low values – about 1 A/m.

Saturation Magnetization

The highest saturation magnetization available in bulk magnetic materials is $\mu_0 Ms \approx 2.43$ T) which is achieved in an iron-cobalt alloy containing 35% cobalt. The possible values of saturation magnetization then range downward continuously to effectively zero. There has been little progress in improving the range of saturation magnetization of materials for about 100 years.

Barkhausen Effect

The Barkhausen effect comprises a series of discontinuous changes in magnetization (either single juamp), sometimes known as 'Barkhausen jumps', that are caused by the sudden motion of domain walls or by the sudden rotation of magnetization within a domain.

These can be observed along the magnetization curve provided sufficient resolution is used in measuring changes in flux density or magnetization.

4.1. Effect of Composition. Properties Relevant for Applications

In fact studies of magnetic properties of amorphous glass coated microwires started as far back as the 70's [84], but recently these tiny glass-coated ferromagnetic wires has been re-discovered mainly due to a number of unusual magnetic properties and their potential applications [2-6].

Generally magneic properties and overall shape of hysteresis loops of amorphous microwires depnends on composition of the metallic nucleus as well as on composition and thickness of the glass coating. This can be illustrated by Figure 6, where the hysteresis loops of 3 main groups of amorphous microwires (Fe-rich, Co-rich and Co-Fe-rich with positive, negative and vanishing magnetostriction constant respectively) and by Figure 7 where hysteresis loops of amorphous microwires with the same composition ($Co_{67}Fe_{3.85}Ni_{1.45}B_{11.5}Si_{14.5}Mo_{1.7}$ with vanishing magnetostriction constant) but with different ρ–ratio defined as d/D (d- metallic nucleus diameter, D- total microwire diameter) are shown. As can be appreciated from Figure 6, shape of hysteresis loops changes from rectangular typical for amorphous Fe-rich compositions to inclined typical for Co-rich compositions. Microwires with vanishing magnetostriction constant exhibit quite soft magnetic properties. Finally Co-rich compositions exhibit almost unhysteretic loops with high enough magnetic anisotropy field (Figure 6). In the case of nearly-zero magnetostrictive $Co_{67}Fe_{3.85}Ni_{1.45}B_{11.5}Si_{14.5}Mo_{1.7}$ microwires, all samples exhibited inclined M(H) loops with extremely low coercivities (up to 4 A/m) but small magnetic anistropy fields, H_k. Magnetic anisotropy field, H_k, increases with increasing the ρ–ratio.

Such strong dependence of the hysteresis loops on these parameters should be attributed to the magnetoelastic energy given by

$$K_{me} \approx 3/2\ \lambda_s \sigma_i, \tag{14}$$

where λ_s is the saturation magnetostriction and σ_i is the internal stress. The magnetostriction constant depends mostly on the chemical composition and is vanishing in amorphous Fe-Co based alloys with Co/Fe ≈70/5 [2]. On the other hand, the estimated values of the internal stresses in these glass coated microwires arising from the difference in the thermal expansion coefficients of metallic nucleus and glass coating are of the order of 100-1000 MPa, depending strongly on the ratio between the glass coating thickness and metallic core diameter [2-5, 85], increasing with the increasing of the glass coating thickness. Such large internal stresses give rise to a drastic change of the magnetoelastic energy, K_{me}, even for small changes of the glass-coating thickness at fixed metallic core diameter. Additionally, such a change of the ρ–ratio should be related to the change of the magnetostriction constant with applied stress [2]:

$$\lambda_s = (\mu_o M_s/3)(dH_k/d\sigma), \tag{15}$$

where $\mu_o M_s$ is the saturation magnetization.

This allows tailoring magnetic properties of glass-coated microwires through the change of magnetic anisotropy by tailoring the internal stresses (by using thermal treatment, chemical etching, etc.) and chamical composition of the metallic nucleus.

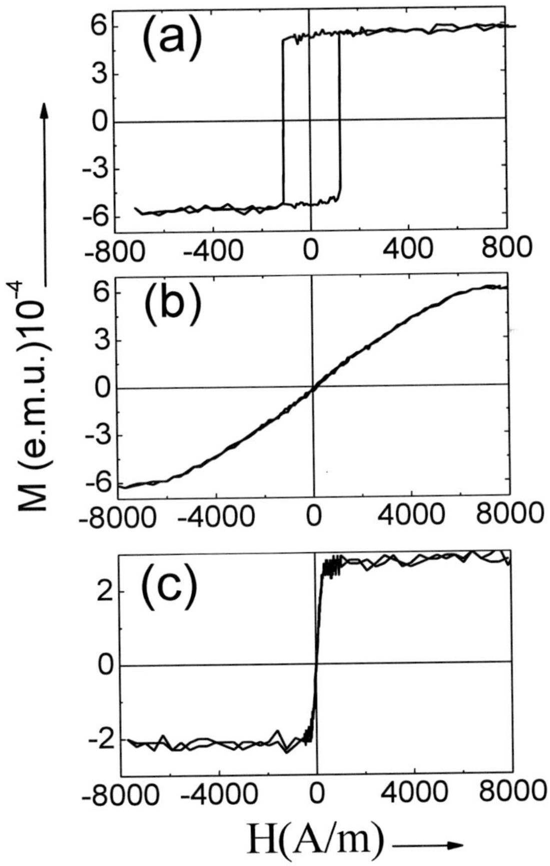

Reprinted with permission from [55], A. Zhukov et al., J. Mater. Res. 15, 2107 (2000). © 2000, Materials Research Society. (figure 1).

Figure 6. Hysteresis loops of thin glass coated Fe-rich (a), Fe-Co-rich (b) and Co-rich (c) microwires.

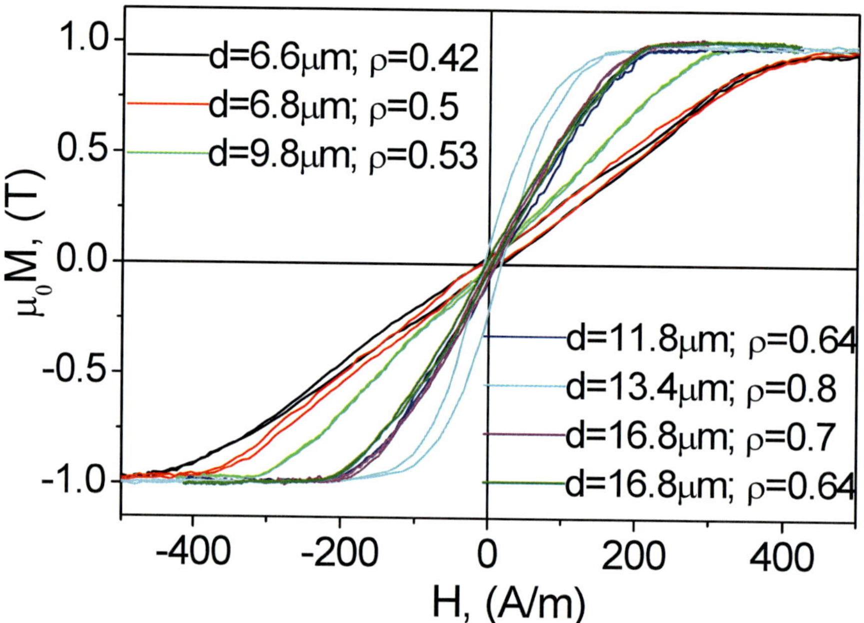

Reprinted with permission from [241] A. Zhukov et al., "Recent advances in studies of magnetically soft amorphous microwires" J. Magn. Magn. Mater. 321, pp.822-825 (2009), Copyright Elsevier B.V (2008) (figure 1).

Figure 7. Hysteresis loops of $Co_{67.1}Fe_{3.8}Ni_{1.4}Si_{14.5}B_{11.5}Mo_{1.7}$ microwires with different geometry.

Main attention of researchers in this field has been paid to the studies of samples with positive magnetostriction constant (Fe-rich compositions) exhibiting perfectly rectangular hysteresis loop (so-called magnetic bistability) and to nearly-zero magnetostriction compositions with good magnetic softness (Co-rich compositions). This is because soft magnetic microwires, exhibiting high magnetic permeability and also GMI effect are quite useful for applications in magnetic field sensors [2-5, 7,8, 47, 48] . On the other hand microwires with rectangular hysteresis loop can be used for magnetic sensors for magnetic surveillance and for magnetic tags [2]. It must be noted again that magnetic behaviour of each group of the microwires dramatically determined by the strong and complex internal stresses acting on the metallic nucleus owing to the glass- coating. As a confirmation, investigations on the progressive elimination of the glass insulating coating by chemical etching on magnetic properties [86,87] showed considerable change of the soft magnetic behaviour as the thickness of glass coating has been diminished.

4.2. Magnetic Bistability. Switching Field Fluctuations. Fast Domain Wall Propagation

Magnetic bistability exhibited by amorphous glass-coated microwires with positive magnetostriction constant is one of the most outstanding properties of these materials. Magnetic bistability is associated with perfectly rectangular hysteresis loop exhibited by microwires related with axial orientation of magnetization with the most part of the metallic nucleus. Most of experimental results are explained by the existing of the inner axially magnetized core inside the metallic nucleus [2-6]. It is still not clear if this axially magnetized core is extended within whole the metallic nucleus or just occupies a considerable part of it. There are some experimental results on measurements of saturation magnetization, remanent magnetization and surface domain structure which imply the existence of the outer transversally magnetized shell [88,89]. On the other hand there are some considerations where existence of these outer domains is questioned [90]. It should be noted that direct observation of the domain structure have been performed after glass coating removal, i.e. when the stresses induced by the glass are removed at least partially. On the other hand, consideration on the absence of outer domains was obtained from the calculated values of the remanent magnetization and its comparison with the saturation magnetization values known from the literature. Probably the most reliable data are on difference of remanent magnetization as compared with saturation magnetization measured in series of Fe-rich samples with different geometry (ratio ρ= d/D). Thus it was observed that remanent magnetization decreases decreasing the ratio ρ, i.e. increasing the thickness of the glass coating [88]. This means that the volume fraction of the microwire with transverse magnetization increases increasing the volume of glass coating. Again, few processes can contribute to the increasing of the non-axial magnetization, such as surface defects, closure domains at the wires ends, interfacial layer between the glass and the metallic nucleus, etc. On the other hand considerable GMI effect observed at high frequencies also confirms the existence of the outer layer with transversal easy magnetization direction [91].

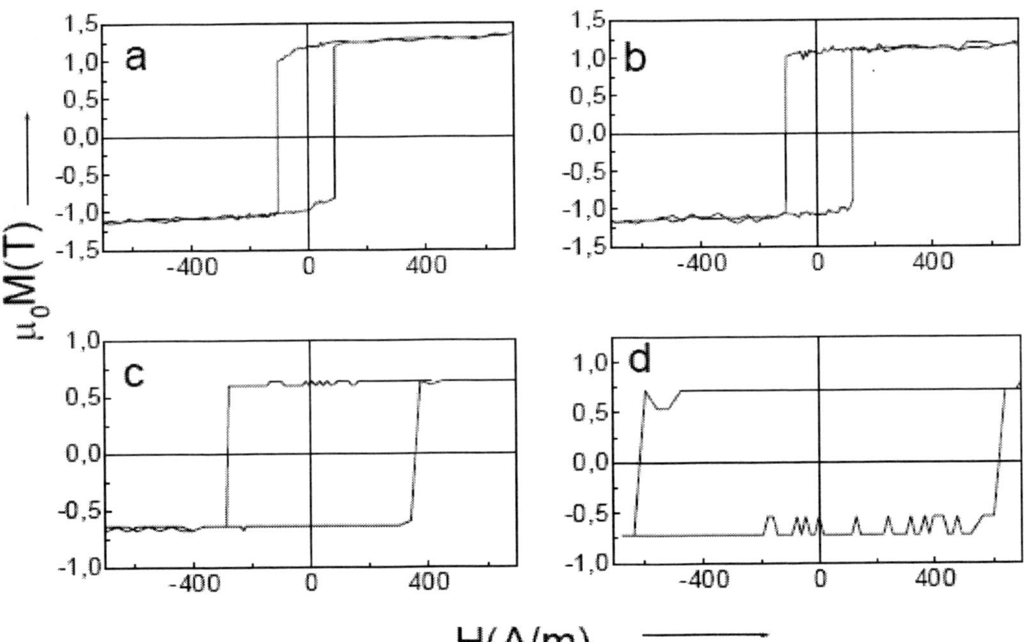

Reprinted with permission from [53] M. Vázquez et al., “Magnetic properties of glass coated amorphous and nanocrystalline microwires”, J. Magn. Magn. Mat. 160, pp 223-228, Copyright (1996) Elsevier Science (figure 3).

Figure 8. Effect of ratio ρ on hysteresis loop of $Fe_{70}B_{15}Si_{10}C_5$ (λ_S >0) microwire (a - ρ=0.63; b - ρ= 0,48; c-ρ =0,26; d-ρ =0,16).

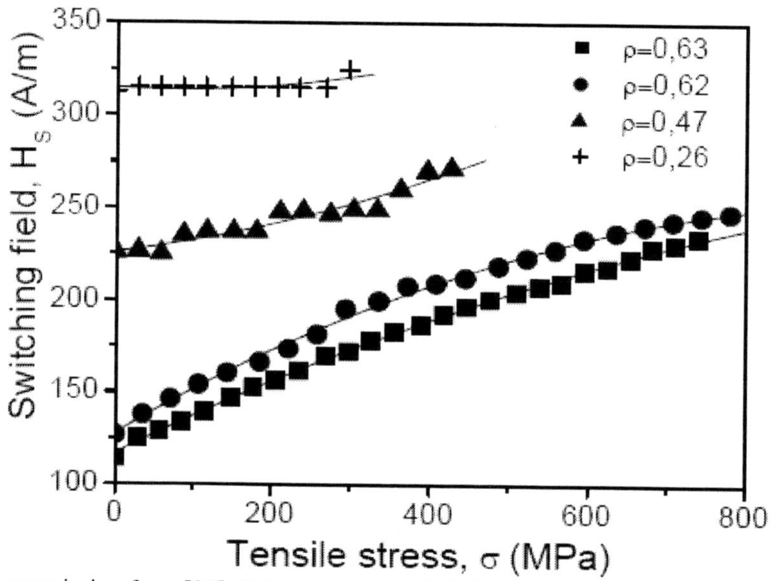

Reprinted with permission from [95], P.Aragoneses et al., "The stress dependence of the switching field in glass-coated amorphous microwires", J. Phys. D: Appl. Phys. 31 (1998) 3040–3045. Copyright (1998) IOP Publishing Ltd. (figure 3).

Figure 9. The tensile stress, σ_a, dependence of the switching field for few Fe-rich microwires with different ρ ratios.

It was assumed that internal stresses, arising during simultaneous fast solidification of metallic nucleus and glass coating, play decisive role in formation of the peculiar domain structure of microwires with positive magnetostriction constant consisting of inner axially magnetized core and outer domains with transverse magnetization [2-5]. Therefore studies of the effect of applied stresses on hysteretic properties are quite useful for understanding the influence of the stresses on magnetic behavior of these microwires. These studies have been performed in several papers for microwires with different ρ-ratio and different compositions [92-96]. In some cases, the effect of glass removal on magnetic properties has been studied too [56, 86, 87, 93, 97,]

As can be observed in Figures 8, 9, even for the same composition of the metallic nucleus quite different hysteretic properties can be observed. The switching field, H_s, that is, the magnetic field required to reverse magnetization (so observing a rectangular hysteresis loop) increases and the remanent magnetization decreases for the same metallic nucleus composition ($Fe_{65}B_{15}Si_{15}C_5$) with decreasing of the geometric ratio ρ. Such increasing of the switching field has been attributed to the increasing of the strength of internal stresses as increasing the glass coating thickness. The switching field, H_s, should be proportional to the energy required to form the domain wall γ involved in the bistable process. The wall energy is related to the magnetoelastic anisotropy and, therefore, to the applied tensile stress as given by [95]:

$$H_s \approx \gamma \propto [3A\,\lambda_S\,(\sigma_a + \sigma_r)/2]^{1/2} / \cos\alpha \qquad (16)$$

where α is the angle between magnetization and axial direction, A is the exchange energy constant, λ_s is the saturation magnetostriction constant and σ_r is the residual tensile stress. Consequently, H_s, must be proportional to $(\sigma_a+\sigma_r)^{1/2}$ for cos $\alpha \approx 1$.On the other hand, it was shown [2-5] that the strength of the internal stresses increases as decreasing the ratio ρ.

Figure 9 shows the tensile stress, σ_a, dependence of the switching field for few Fe-rich microwires with different ρ–ratios. The switching field, H_s, increases monotonously with the applied stress for the Fe-rich microwires with different ρ-ratio.

Applied stress dependence of H_s is rather different in microwires as-compared with in-water quenched amorphous wires. Thus, "conventional" amorphous wires usually exhibit a minimum on $H_s(\sigma_a)$ dependence at low values of σ_a (as observed for example in [95, 98]). This minimum does not observed on H_s (σ_a) dependencies for Fe-rich microwires. This difference can be explained according to eq. (16) from the existence of elevated quenched-in internal stresses arising mainly due to the glass coating. Another experimental detail is that $H_s(\sigma_a)$ dependence for Fe-rich compositions is rather weak for smallest ρ-ratios (see Figure 9). It confirms the existence of additional axial stresses arising from the glass coating. Generally, as predicted by eq.16 and observed in Figure 9 as well as for "conventional" amorphous wires at elevated σ_a, $H_s(\sigma)$ follows $\sigma^{1/2}$ dependence, where σ is the total (applied and internal) stresses, therefore an existence of internal stresses makes experimental $H_s(\sigma)$ dependence weaker for small ρ–ratios [95].

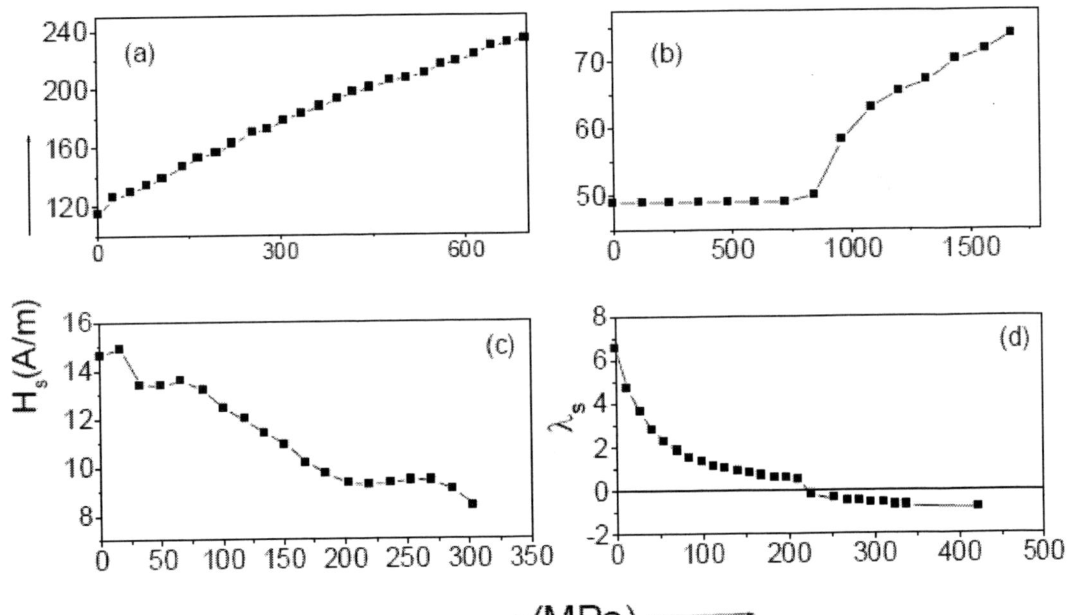

Reprinted with permission from [100] A. Zhukov "Glass-coated magnetic microwires for technical applications", J. Magn. Magn. Mat. 242-245, (2002) pp 216-223, Copyright (2002), with permission from Elsevier Science (figure 9).

Figure 10. Effect of microwire composition on $H_c(\sigma)$ dependence: $Fe_{65}B_{15}Si_{15}C_5$ (a), $(Co_{1-x}Mn_x)_{75}Si_{10}B_{15}$ with x =0,11 (b), x = 0,09 (c), $\lambda_s(\sigma)$ dependence for x = 0,09 (d).

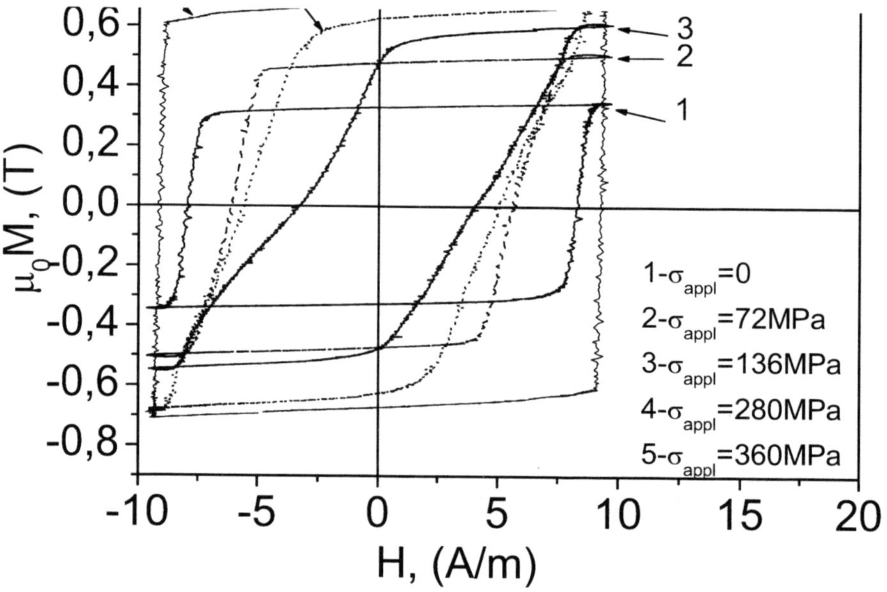

Reprinted with permission from [105], "Effect of stress applied on the magnetization profile of Fe–Si–B amorphous wire". V. Zhukova et al., *J.* of Appl.Phys. 93, 10 (2003). Copyright © 2003 American Institute of Physics (AIP). (figure 1).

Figure 11. Stress dependence of the hysteresis loops of 8 cm long $Fe_{77.5}B_{15}Si_{7.5}$ amorphous wire with 125μm diameter, measured by a long pick-up coil.

It is worth mentioning that interesting influence of applied tensile stress on hysteretic properties can be observed in glass-coated amorphous microwires if the magnetostriction constant is positive but small. These dependences have been studied in Fe-Co and Fe-Mn rich microwires. The behavior of coercivity and remanence with applied tensile stress depend strongly on the microwires composition and geometrical characteristics exhibiting a quite different character for Fe- and Co-rich compositions (Figure 10). Generally, a proportionality of the switching field, H_s, to the squire root of the applied stresses has been observed in Fe-rich compositions, as predicted by eq. (16) (Figure 10 a). $H_s(\sigma)$ dependence in Co-rich $(Co_{1-x}Mn_x)_{75}Si_{10}B_{15}$ (x = 0.08 - 0.11) microwires with small positive magnetostriction constant is quite different from Fe-rich microwires [81, 94-96]. Besides, the evolution of $H_s(\sigma)$ depends also on the composition.

In fact, some increase of the coercivity H_c with σ after initial roughly independent behavior has been observed for x=0.11 (Figure 10b). The sample $(Co_{0.9}Mn_{0.1})_{75}Si_{10}B_{15}$ with initially rectangular hysteresis loop shows a decrease of the coercivity and the remanent magnetization with σ and, finally, lost the magnetic bistability after application of certain σ (Figure 10c).

The estimations of the magnetostriction constant were based on the variation of the initial magnetic permeability with tension in such microwires [81]. According to this method, the magnetostriction constant, λ_s, can be evaluated from the relation:

$$\lambda_s = -(\mu_o M_s^2/3)d(\chi_i^{-1})/d\sigma \qquad (17)$$

where M_s is the saturation magnetization, χ_i - initial magnetic susceptibility and σ– applied tensile stress.

The dependence of λ_s **(σ)** for the sample with 0.09 is shown in the figure 10d. It should be noted that the observed evolution have mostly illustrative character, since the level of the magnetic response from a single microwire is quite small. Sample with x=0.09 exhibiting positive magnetostriction constant shows also a decrease of the λ_s with σ and even changes its sign at certain σ (around **200** MPa) .

The stress dependence of the magnetostriction constant, λ_s, (Figure 10d) allows to ascribe positive values of the order of 10^{-7} for the magnetostriction constant in the sample with x $\geq$ 0.09. Such as it was found in other amorphous materials, the application of tensile stresses results in a decrease of the λ_s in case of positive magnetostriction constant, that even can change the sign of λ_s in the sample with x=0.09 when σ reaches some certain value (see figure 10d). In fact it was experimentally found in nearly-zero magnetostrictive amorphous ribbons [58] that

$$\lambda(\sigma)= \lambda(0)-B\sigma \qquad (18)$$

where $\lambda(0)$ is the saturation magnetostriction constant without applied stresses and B is a positive coefficient of the order of 10^{-10} $(MPa)^{-1}$. For magnetostriction $\lambda(0)$ of about $1x10^{-7}$ a stress of around 10^3 MPa needs to be applied to reach compensation of magnetostriction [87,100, 101]. In fact, internal stresses induced during the preparation can reach that values which indicates that high quenched-in internal stresses seems to be able to change effectively the value of the magnetostriction constant [87].

One of the main technological interests for utilisation of amorphous microwires is related with Large and single Barkhausen Jump (LBJ) observed above some value of applied magnetic field, called switching field between two stable remanent states[2].

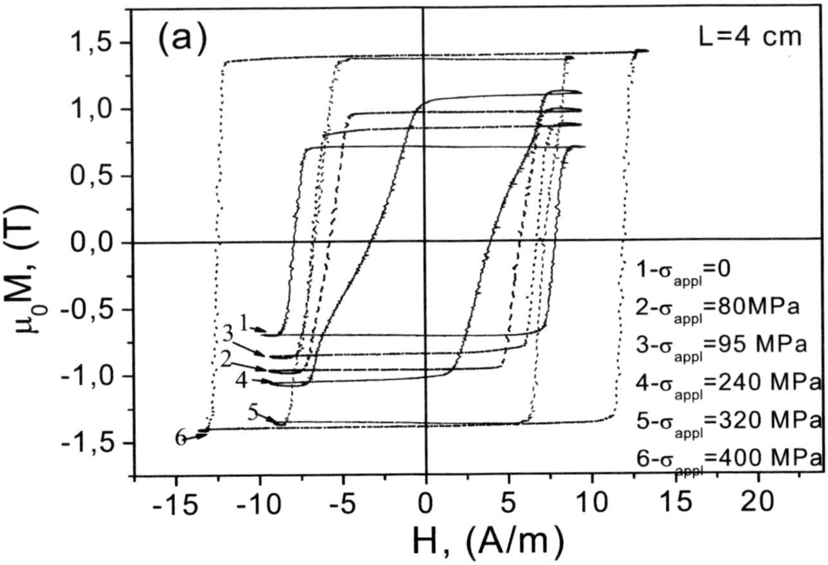

Figure 12. Continued on next page.

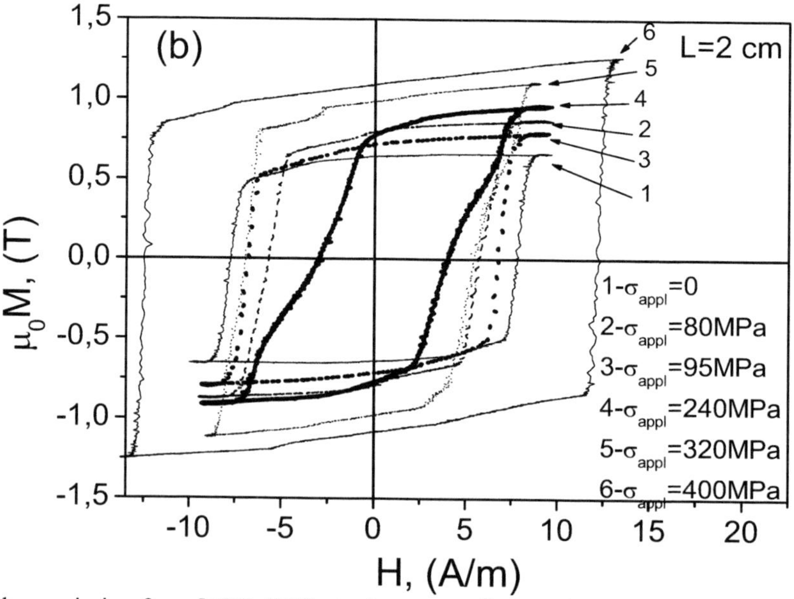

Reprinted with permission from [105], "Effect of stress applied on the magnetization profile of Fe–Si–B amorphous wire". V. Zhukova et al., J. of Appl..Phys. 93, 10 (2003). Copyright © 2003 American Institute of Physics (AIP) (figure 2).

Figure 12. Effect of stress applied on hysteresis loops of 8 cm long $Fe_{77.5}B_{15}Si_{7.5}$ amorphous wire, measured by a short movable coil placed at *L*= 4cm (a) and 2 cm (b).

It is worth mentioning that appearance of Large and single Barkhausen jump takes place under magnetic field above some critical value (denominated as switching field) and also if the sample length is above some critical value denominated also as critical length. As described above, switching field depends on magnetoelastic energy determined by the strength of the internal stresses, applied stresses and magnetostriction constant determined by the chemical composition of ferromagnetic metallic nucleus [2-5].

Regarding the critical length, detailed studies of the ferromagnetic wire diameter on magnetization profile and size of the edge closure domains have been performed in [13]. Particularly, critical length, l_c, for magnetic bistability in conventional Fe-rich samples (120 μm in diameter) is about 7 cm. This critical length depends on saturation magnetization, magnetoelastic energy, domain structure, magnetostatic energy [13, 102-104]. Thus, in Co-rich conventional amorphous wires (120 μm in diameter) such critical length is about 4 cm [102, 103]. Below such critical length hysteresis loop loses its squared shape.

The magnetostatic energy depends on the demagnetizing field, H_d, expressed as:

$$H_d=NM_s \tag{19}$$

where N is the demagnetizing factor given for the case of long cylinder with length, ℓ, and diameter, *D*, as:

$$N=4\pi(ln(2\ell/D)-1](D/\ell)^2) \tag{20}$$

On the other hand the shape of the hysteresis loop of wires with a given length depends on the applied stress. As an example, Figure 11 shows the effect of the applied stress on the hysteresis loop of the 8 cm long sample measured by the long secondary coil. It is worth to mention that the initially rectangular shape of the hysteresis loop is lost under the certain applied stress (around 70 MPa) but finally it recovers under stronger applied stress of around 320 MPa.

Similar experiments has been done by using short movable pick-up coil which permitted to measure local hysteresis loops as well as to obtain the magnetization profile and its dependence on applied stress. Such dependence measured in the center of the 8 cm long wire, i.e. at the position of the movable pick-up coil of L=4 cm and at L=2 cm are presented in Figure 12 (a, b) respectively. At L=4 cm the hysteresis loop measured for σ_{appl} <60 MPa and σ_{appl} >320 MPa exhibit squared shape, while at L =2 cm all hystsresis loop are not perfectly rectangular.

The remanence, $\mu_0 M_r$, profile has been acquired from the measurements of local hysteresis loops. Such profiles measured for the 8 cm long wire is shown in Figure 13.The remanence profile has roughly symmetric character. The decrease of the remanent magnetization close to the wire ends (variation of $\mu_0 M_r$ with the position of the coil, L) as well as the variation of the penetration depth, ℓ_d, of the closure domains with σ_{appl} are observed in Figure 13. As can be assumed from Figure 13, the penetration depth, ℓ_d first increases and then decreases with σ_{appl}. Below certain sample length, L≈ ℓ_c ≈2 ℓ_d, magnetic bistability disappears because the closure domains penetrate far enough inside the inner axially magnetized core and destroy it.

Consequently the critical length first increases with σ_{appl} but then again decreases and finally magnetic bistability recovers under strong enough applied tensile stress.

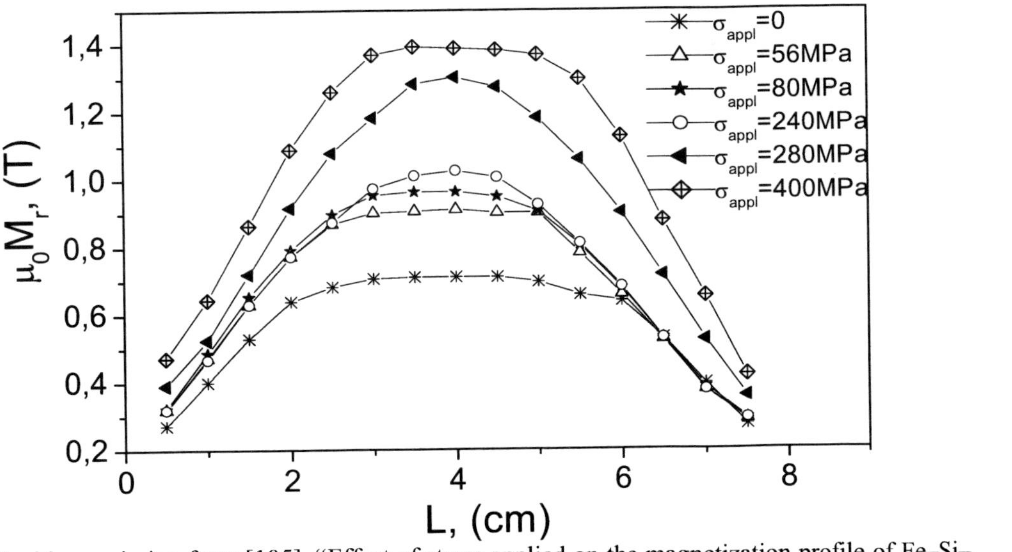

Reprinted with permission from [105], "Effect of stress applied on the magnetization profile of Fe–Si–B amorphous wire". V. Zhukova et al., J. of Appl.Phys. 93, 10 (2003). Copyright © 2003 American Institute of Physics (AIP) (figure 5).

Figure 13. Effect of stress on magnetization profiles measured in $Fe_{77.5}B_{15}Si_{7.5}$ amorphous wire measured for sample length l=8 cm.

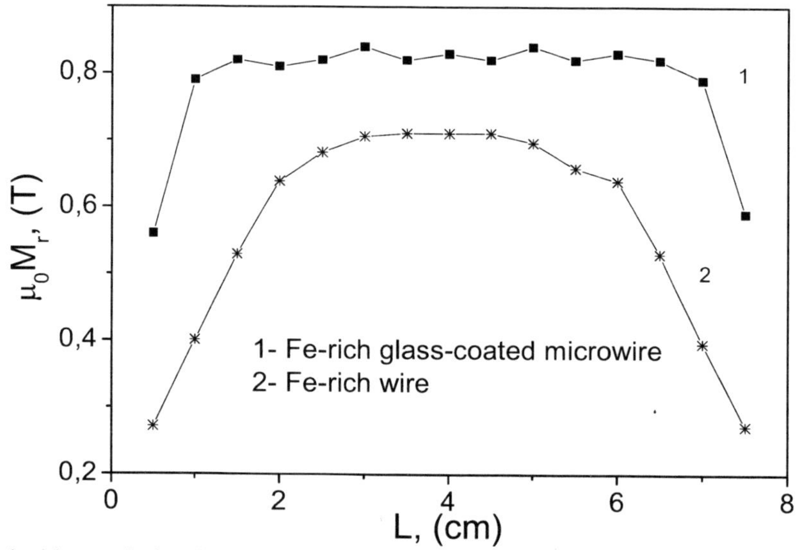

Reprinted with permission from [13] A.P. Zhukov, M. Vazquez, J. Velazquez, H. Chiriac, V. Larin "The remagnetization process in thin and ultra-thin Fe-rich amorphous wires" J. Magn. Magn. Mat. 151(1995) 132-138. Copyright (1995) Elsevier Science (figure 4).

Figure 14. Comparison of the magnetization profiles measured in Fe-rich wire with 125mm diameter and Fe-rich microwire.

General increase of $\mu_0 M_r$ with σ_{appl} should be attributed to the increasing of the volume of axially magnetized area under the effect of applied stresses due to the positive magnetostriction constant. On the other hand in the range $60\ MPa \le \sigma_{appl} \le 300\ MPa$ magnetic bistability disappears and the local hysteresis loops do not show rectangular character even in the central region of 8 cm long wire (see Figure 11). Such behavior is related with the demagnetizing factor of the internal axially magnetized core: the diameter of such single domain core increases which results in increasing of the demagnetizing factor given by eq. (20). This affects the penetration depth of the closure domains, ℓ_d, and in this way the magnetic bistability is destroyed even in the central zone of wire.

Further increase of applied stresses results in further increasing of the remanent magnetization and also in drastic increasing of the coercivity and recovering of magnetic bistability. We assume that the increase of the external tensile stress results in the increase of the domain wall energy,

$$\gamma = 2(AK)^{1/2} \quad (21)$$

where A is the exchange energy and K is the magnetic anisotropy constant, which in the case of amorphous state depends mainly on the magneto-elastic component, K_{me}, given by eq. (14).

Applying external stress the magnetoelastic energy increases and, consequently, the domain wall energy, γ. At such conditions there is a competition between magnetostatic and magnetoelastic energies. Such increasing of the magnetoelastic energy gives rise to the decreasing of the size (and consequently of the penetration depth) of closure domains and therefore results in new decreasing of the critical length and recovering of the magnetic bistability.

Phenomenon of magnetic bistability observed in different families of amorphous wires is certainly quite interesting for the magnetic sensors applications. But, enhanced critical length, ℓ_c, (of the order of few cm) observed in conventional amorphous wires limited these applications. On the other hand, reduction of the metallic nucleus diameter in the case of glass-coated microwires (almost one order lower) results in drastic reduction of the critical length, making them quite attractive for micro-sensor applications. Thus, magnetic bistability for the sample length L= 2 mm has been observed for Fe-rich microwire with metallic nucleus diameter, d, about 10 μm [13]. Remanent magnetization profile measured in Fe-rich microwires exhibit deviations just near the sample ends (see comparison of the magnetization profiles in Figure 14). Strong internal stresses (mostly of axial origin), appearing during the fabrication of glass coated microwires, can also contribute to the reduction of the critical length for magnetic bistability.

One of the extraordinary properties exhibited by amorphous microwires with positive magnetostriction constant is observation of single and large Barkhausen jump, allowing studying details of remagnetization process by relatively simple methods.

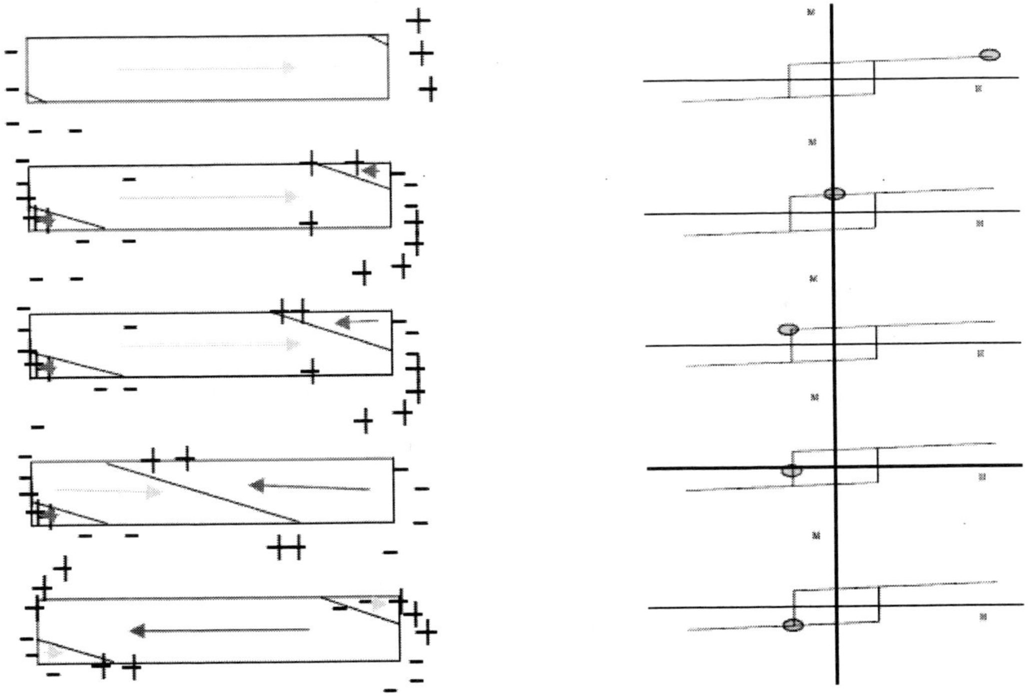

Figure 15. Schematic domain structure of positive magnetostriction microwires and its evolution with applied field. In parallel, the corresponding position at the hysteresis loop is given.

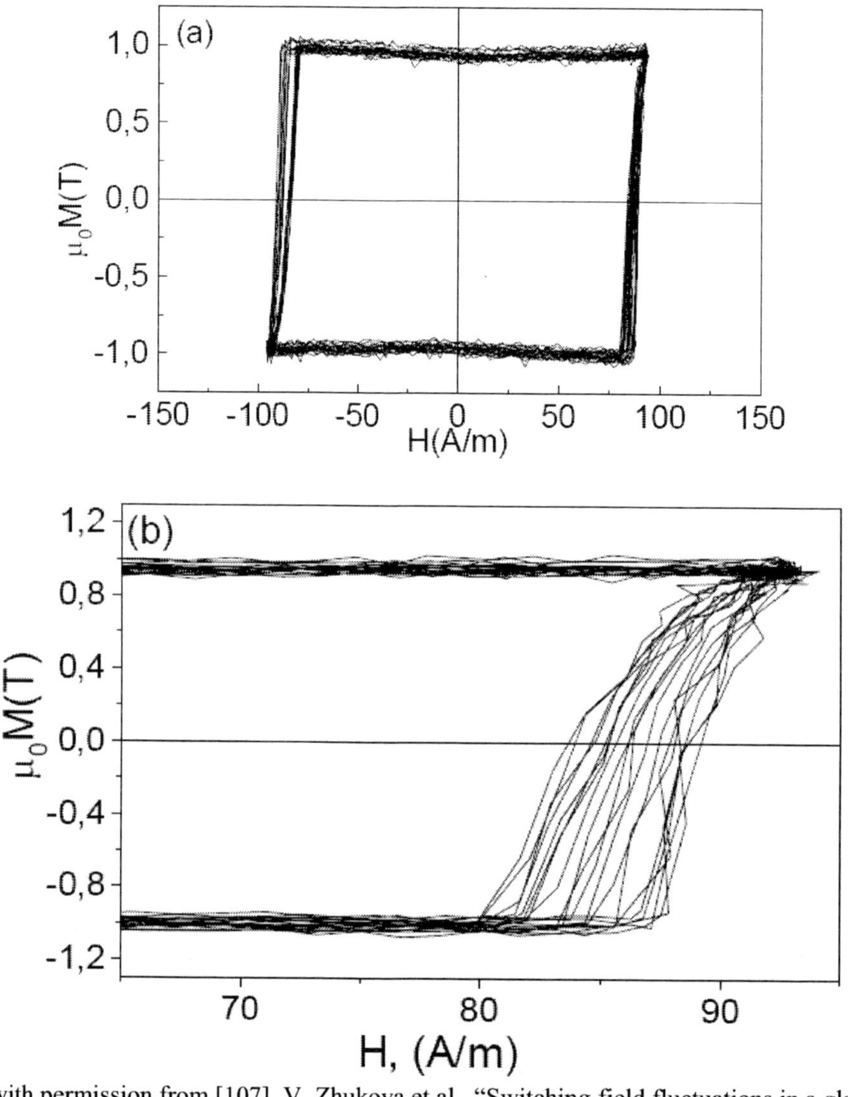

Reprinted with permission from [107], V. Zhukova et al., "Switching field fluctuations in a glass-coated Fe-rich amorphous microwire", *J. Magn. Magn. Mater.* 249, (2002) pp 131-135, Copyright (2002) Elsevier Science (figure 1).

Figure 16. Hysteresis loop of the re-entrant remagnetization process: (a) and switching field fluctuations and (b) in glass coated amorphous $Fe_{65}Si_{15}B_{15}C_5$ microwire.

Such particular magnetization process has been observed previously in different kinds of amorphous wires [48, 49, 105-106] and also in iron whiskers [107], amorphous Co-rich ribbons after special heat treatment [108-111]. Sharp magnetization jump and sharp electrical pulses related with the magnetization switching during LBJ have been used in different kinds of magnetic sensors, such as magnetic markers and magnetoelastic sensors [8,9].

Nevertheless, the use of these materials is limited, due to intrinsic switching field fluctuations observed in all these materials [106-110]. On the other hand, studies of the switching field fluctuations allow understanding better remagnetization process of magnetically bistable magnetic microwires.

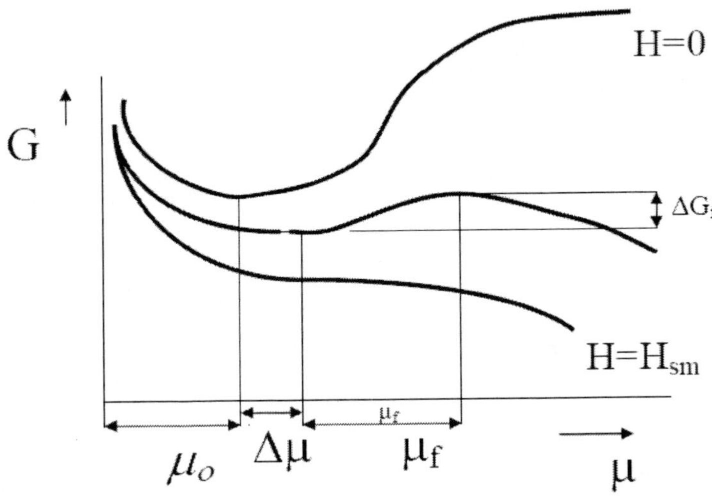

Reprinted with permission from [107], V. Zhukova et al., "Switching field fluctuations in a glass-coated Fe-rich amorphous microwire", *J. Magn. Magn. Mater.* 249, (2002) pp 131-135, Copyright (2002) Elsevier Science (figure 2).

Figure 17. Schematic representation of the model: dependence of the free energy, G; of reversed domain on its magnetic moment, μ; is plotted for different values of the applied magnetic field.

4.2.1. Switching Filed Fluctuations

In fact switching filed fluctuations have been observed in most of magnetic materials exhibiting rectangular hysteresis loop when the remganetization process runs through single and large Barkhausen jump [106-111]. Particularly, phenomenological description of switching field fluctuations has been first developed for amorphous ribbons [108-110] and later extended to glass-coated microwires [106, 111-113]

Such switching field fluctuation results in spontaneous experimental spread in the value of the switching field taking such value in different re-magnetizing cycles [106-113]. These switching field fluctuations can deteriorate the performance of the magnetic marker based on a number of encoding combinations because of the overlapping of the switching field values from different microwires and of the magnetoelastic sensor making its work unreliable.

Phenomenological model explaining the onset of switching field fluctuations from the effect of thermal fluctuations on the remagnetization process have been developed for the case of amorphous ribbons. Anyway, the remagnetization process of amorphous ribbons seems to be rather complex in order to be explained by such simplified model. In ribbons, this process takes place mainly by displacements of one or two domain walls through the cross section [109]. In contrast, the remagnetization process of wires is much clearer. Roughly single domain structure is considered for the inner axially magnetized core. Reversed domains were observed at the ends due to the demagnetizing field. At certain applied magnetic field a depinning and/or nucleation of reversed domains at the ends of wires have been observed by direct domain wall observation [106, 112] (see Figure 15). In this sense, the re-magnetization behaviour of wires is simpler and nucleation sites are clearly identified being located at the wire ends.

A spontaneous spread in the switching field values during the re-entrant magnetization with a random distribution of the switching field values, ΔH_s, around some average value H_s has been first measured in $Fe_{65}B_{15}Si_{15}C_5$ microwire (see Figure 16). Detailed observation of the area close to the switching field permits to study the switching field distribution. Observed experimental spread is quite significant (~10 A/m) being around 10% of the mean switching field value (~85 A/m). It is remarkable that a previously developed phenomenological model attributed these switching field fluctuations to the thermal fluctuations of the magnetic moments of the magnetization nucleus [108-110].

This model assumes that the magnetic moment of the reversed domain is a function of the applied magnetic field. At any applied field, the reversed domain obtains an equilibrium size, i.e. equilibrium magnetic moment, μ. For instance, at H=0 and $\mu=\mu_0$ the energy of the reversed domain has a minimum. In fact the existence of the reversed domains at the wire ends and it growth with applied field before switching has been confirmed by direct domains observation and magnetization profiles measurements [13, 115].

Described features are shown schematically in the Figure 17 where the dependence of the free energy, G, of reversed domain on its magnetic moment, μ, is plotted for the different values of applied magnetic field. As can be deduced from the Figure 17, in principle the remagnetization begins at $H=H_{sm}$ when the energetic barrier disappears under the effect of applied magnetic field. However, the remagnetization can occur even at slightly lower magnetic field ($H=H_{sm}-\Delta H$) under the effect of thermal fluctuations. In such case the energetic barrier, ΔG, should be overcome by the thermal fluctuation of the magnetic moment, μ_f. In this case, re-entrant hysteresis loop will exhibit switching field fluctuations.

This model, initially developed for amorphous ribbons with magnetic bistability, predicts the shape of the switching field distribution: dependence of the logarithm of the probability density, dw, to observe large Barkhausen jump in the range between ΔH and ΔH+d(ΔH) on reduced switching field fluctuation $(\Delta h)^{3/2}$.

$$\ln\left[\frac{dw}{d(\Delta h)}\right] = \beta - \alpha(\Delta h)^{3/2} \tag{22}$$

where

$$\alpha = \frac{8\chi_0 H_{sm}^2}{3kT} \tag{23}$$

$$\Delta h = \frac{\Delta H}{H_{sm}} \tag{24}$$

$$\beta = \frac{2}{3}\ln\alpha - \ln 2 - \ln\left[y\left(\alpha^{1/3}\right)\right] \tag{25}$$

$$y\left(\alpha^{1/3}\right)=\int_0^{\alpha^{1/3}} x\exp(x^3)dx \qquad (26)$$

and χ_0 is the susceptibility of the reversed domain at zero applied magnetic field:

$$\chi_0=\left(\frac{\partial\mu}{\partial H}\right)_{H=0} \qquad (27)$$

As follows from (22)-(26), this model predicts a linear dependence of the logarithm of the probability density, dw/d(Δh), on $(\Delta h)^{3/2}$.

In order to compare the experimental results with predictions of the model, the switching field distribution has been measured in $Fe_{65}B_{15}Si_{15}C_5$ microwire. Switching field histogram is introduced as normalized number of LBJ, N_k, per field interval determined as $\delta h = 1/k$, where k is the number of such intervals within the range of the magnetic fields where the switching field fluctuations are observed:

$$U_k=N_k/\delta h\bullet N \qquad (28)$$

where N is the total number of observed hysteresis cycles. Obtained histogram of measured switching field distribution has an asymmetric shape similar to that observed for amorphous ribbons in [108-110].

Density of probability to observe the switching field fluctuation in the range Δh - Δh+d(Δh) is expressed by:

$$B\nu(\Delta h)=\frac{dw}{d(\Delta h)} \qquad (29)$$

If the value of Δh is small enough, it is possible to approximate the function $\nu_k = \nu(\Delta h=\kappa\delta\eta)$ as:

$$\nu_k \approx \nu\,[(k-1)\delta h+\delta h/2] \qquad (30)$$

In this case the probability to observe the switching field fluctuation in the field interval $(k-1)\delta h<\delta h<k\delta h$ is expressed as:

$$dw \approx B\,\nu_k\delta h \qquad (31)$$

Leading to the expression for the probability density:

$$\frac{dw}{d(\Delta h)} \cong B v_k = \frac{N_k}{\delta h \sum_{i=k}^{k} N_i} \left[\sum_{k=1}^{k_m} \frac{N_k}{\sum_{i=k}^{k_m} N_i} \right]^{-1} \tag{32}$$

Such probability takes into account the peculiarities of our experiment, such as that the magnetic field always changes during the experiment increasing from zero till H_{sm}.

Consequently, experimental dependencies re-plotted taking into account (32) are shown in Figure 18. As can be observed from Figure 18, this dependence fits quite well by the linear dependence, as it was predicted by the equation (22).

On the other hand, more detailed studies of the switching filed distribution performed at different temperatures, as well as in microwires with different chemical composition of the metallic nucleus (and consequently with different magnetostriction constant) exhibit some disagreement with this model [106,112,113]

Temperature dependence of the switching field fluctuations have been measured for few microwires (see compositions in Table 1)

The temperature dependence of the switching field distribution of $Fe_{77.5}Si_{7.5}B_{15}$ microwire is given in Figure 19. At low temperatures, the switching field distribution is wide and average switching field is elevated as –compared with room temperature value. Increasing the temperature, the distribution width decreases and whole distribution shifts to lower fields. At higher temperatures, the average switching filed (which corresponds to the distribution position) almost does not change with temperature as well as the distribution width.

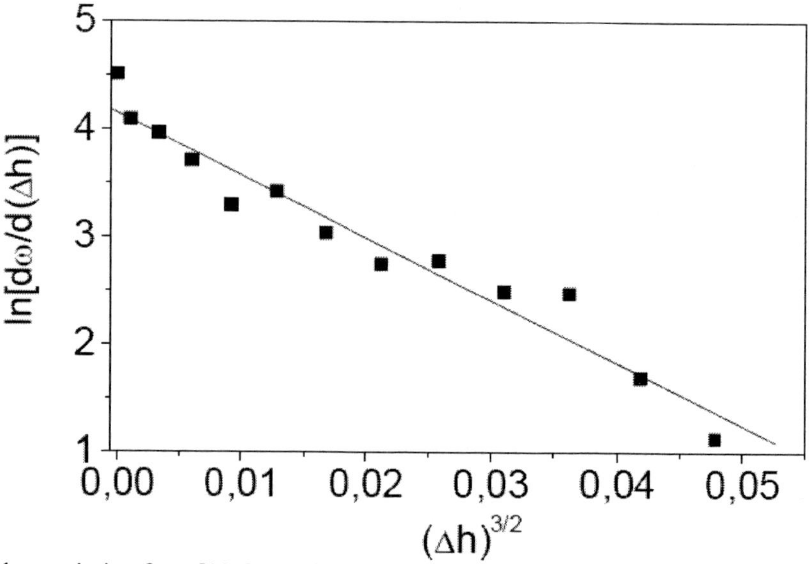

Reprinted with permission from [107], V. Zhukova et al., "Switching field fluctuations in a glass-coated Fe-rich amorphous microwire", *J. Magn. Magn. Mater.* 249, (2002) pp 131-135, Copyright (2002) Elsevier Science (figure 4).

Figure 18. Logarithm of the probability of switching field fluctuations versus $(\Delta h)^{3/2}$.

Table.1.Composition, dimensions and magnetostriction of studied microwires [112,113]

Composition	diameter of metallic core (μm)	Total diameter (μm)	Magnetostriction (x10^{-6})
$Fe_{77.5}Si_{7.5}B_{15}$	11	29	25
$Fe_{36}Co_{40}Si_{13}B_{11}$	15	27	15
$Co_{68}Mn_7Si_{15}B_{10}$	8	20	0.5

The temperature dependence of the $Fe_{36}Co_{40}Si_{13}B_{11}$ microwire is given in the figure 20. The distribution is wide at low temperature, but the temperature dependence of the switching field distribution width is weaker than in the case of FeSiB microwire. Additionally, temperature dependence of the switching field is quite different from $Fe_{77.5}Si_{7.5}B_{15}$: temperature dependence at low temperature is weaker, but it is more significant at higher temperatures.

The switching field distribution of $Co_{68}Mn_7Si_{15}B_{10}$ microwire measured at different temperatures is given in figure 21. Here we can see the strong effect of the decrease of the coercivity at low temperatures, which continues up to 350K. Above this temperature, the microwire of the given composition looses its bistability and the study of the switching field distribution is no more possible. As in the previous samples, the distribution width decreases with the temperature.

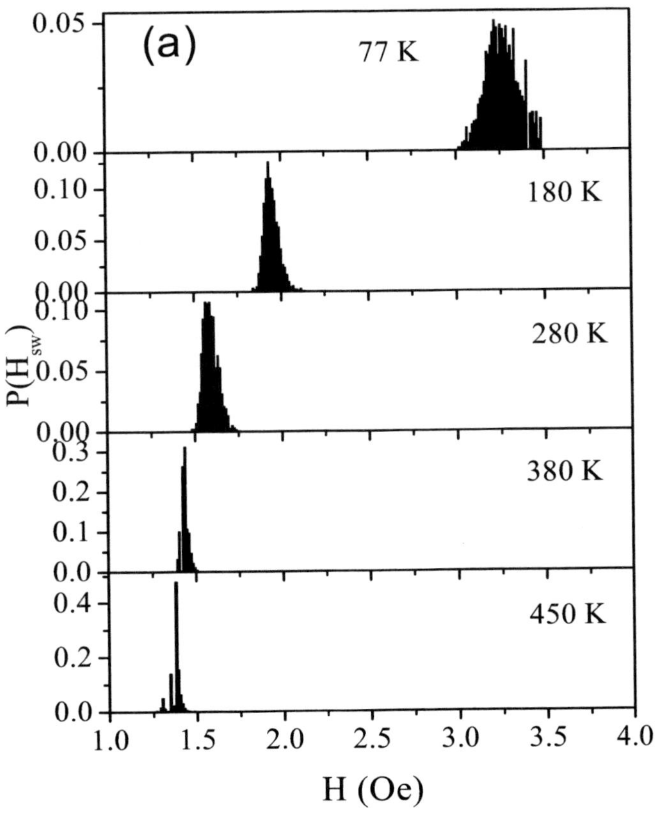

Figure 19. Continued on next page.

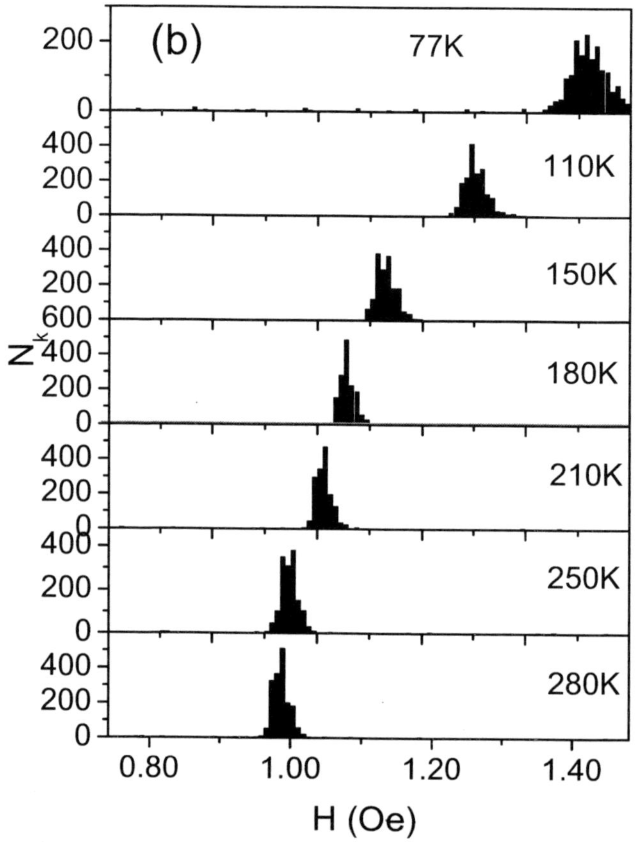

(a) Reprinted with permission from [113] R. Varga et al., "Magnetization processes in thin magnetic wires" *J. Magn. Magn. Mater.* 300 (2006) e305–e310. Copyright (2006), with permission from Elsevier Science. (figure 4)(b) Reprinted with permission from [113],R. Varga et al., "Temperature dependence of the switching field and its distribution function in Fe-based bistable microwires", Appl. Phys. Let. V83, 13, 2620 (2003). Copyright American Institute of Physics (2003)(figure 1).

Figure 19. The switching field distribution of the FeSiB microwire measured at different temperatures.

Without the action of the external field, the free energy *G* of the closure domain wall is given by its potential (*W*) (see figure 17), which is given on the macroscopic scale by the sum of the magnetostatic, magnetoelastic etc. interaction. The external magnetic field, *H*, changes the shape of the free energy *G* of the closure domain wall. The free energy *G(x)* of the closure domain wall is then given by [108-110]:

$$G(x) = W(x) - 2\mu_0 M_s H A x, \qquad (33)$$

where μ_0 is the permeability of vacuum, M_s is the saturation magnetisation, *A* is the area of the rigid segment of the considered DW. If the external field *H* reaches the value of the switching field H_{sw}, the minimum of total energy disappear and the DW is forced to jump into the new position in one Large Barkhausen Jump (see figure 17). The switching field is given by the relation [106, 108-110]:

$$H_{sw} = \left.\frac{dW}{dx}\right|_{max} . \tag{34}$$

In the given model, the energy barrier ΔG given in the figure 17 could be overcome by the thermal activation or by tunnelling even in the case, where $H < Hsw$, [106, 108-110]. In that case, we can found the switching field fluctuation and the distribution of the switching fields will be measured. The probability w to overcome the free energy barrier is given by:

$$w = A.\exp(-\Delta G/kT) \tag{35}$$

where A is a constant and the dependence of the energy barrier ΔG on the reduced external field Δh ($\Delta h=(H_{sw}-H)/H_{sw}$) can be described by [106,108-110]:

$$\Delta G = (8/3)\chi_0 H_{sw}^{2}(\Delta h)^{3/2}. \tag{36}$$

Here χ_0 is the susceptibility of the reversed domain at zero external field.

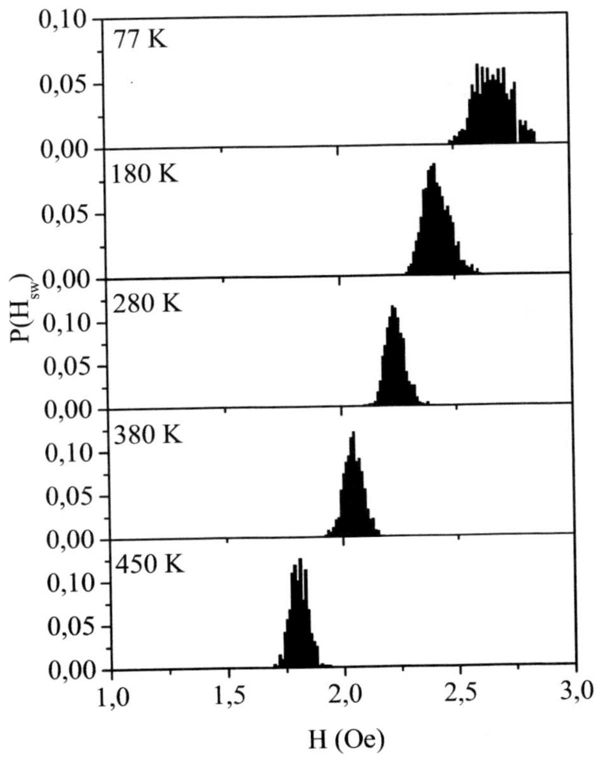

Reprinted with permission from [114], R. Varga et al., "Distribution and temperature dependence of switching field in bistable magnetic amorphous microwires" in Recent Res. Devel. Non-Crystalline Solids, 3 (2003) pp.85-95, ISBN: 81-7895-090-1, Publisher: Transworld Research Network, Trivandrum, India (2003) (figure 3).

Figure 20. The switching field distribution of the FeCoSiB microwire measured at different temperatures.

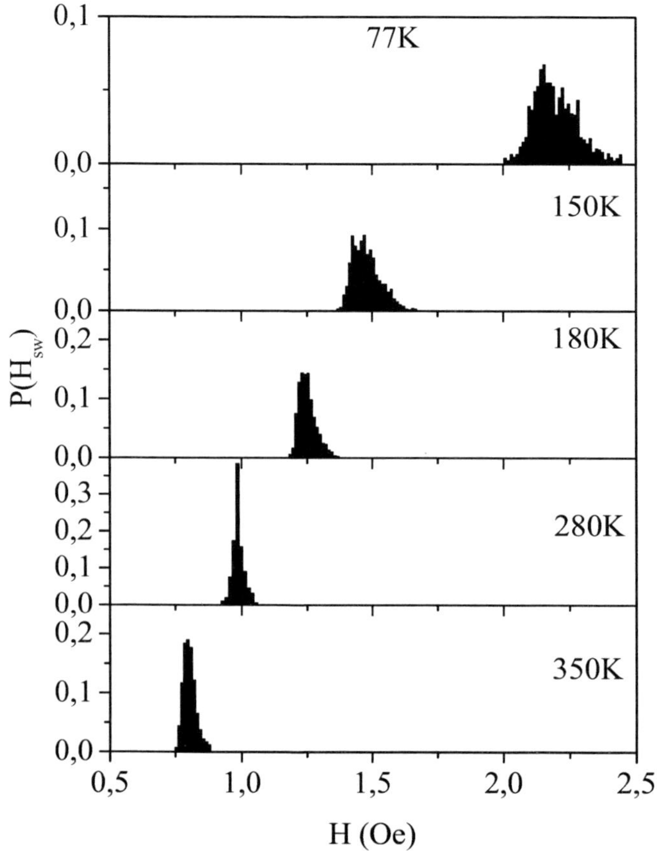

Reprinted with permission from [114], R. Varga et al., "Switching-field distribution in amorphous magnetic bistable microwires" Phys. Rew. B 70, 024402 (2004), Copyright © (2004) The American Physical Society (figure 2);

Figure 21. The switching field distribution of the CoMnSiB microwire measured at different temperatures.

As follows from eq.22, the linear dependence of the logarithm of the probability density $dw/d(\Delta h)$ on $(\Delta h)^{3/2}$ is predicted by this model.

As can be seen from figure 22, linear dependence has not been observed in the case of microwires. But, observed dependence can be described as the sum of two linear dependencies. This can be explained in term of the sum of two different potentials instead the one given in eq.22. The two potentials can be associated with the magnetoelastic and stabilisation potentials as was found also from the temperature dependence of the coercivity [112,113, 116]. The magnetoelastic potential comes from the long range interaction and therefore its shape is wide. On the other hand, the potential coming from pinning on atomic scale defects has short-range origin. Therefore, its shape is more narrow than of magnetoelastic one (figure 23 a,b.). The free energy minimum of the domain wall is then given by the sum magnetoelastic potential and potential raising from pinning on atomic scale defects which results in the two different slopes α_1 and α_2 (eq.22).

This assumption has been confirmed experimentally [112-114]. In order to distinguish between the role of the particular contributions, the measurements under the applied stress (to moderate the magnetoelastic contribution) as well as at a different frequencies (to moderate the time dependent structural relaxation) were performed. At certain conditions (low frequency regime) almost linear dependence of $ln(dw/d(\Delta H))$ on the reduced field $\Delta h^{3/2}$ has been observed (figure 24). This can be explained because at low frequencies (20Hz), the structural relaxation has enough time to stabilize the domain wall. Increasing the frequency of the measurement the measuring time decreases and there is not enough time for the developing of the structural relaxation. In such case, another minimum appears in the domain wall free energy at intermediate fields H_2 (figure 23). Now, the probability of switching p is given as a product of two probabilities p_1,p_2 that the corresponding minimum will be overcomed ($p=p_1*p_2$) therefore, the slope of the probability density vs. reduced field decreases. At the field H_3 (figure 23) close to the switching field ($\Delta h^{3/2} \rightarrow 0$), the second minimum disappears and the slope of the dependence of $ln(dp/d(\Delta H))$ on the reduced field $\Delta h^{3/2}$ increases.

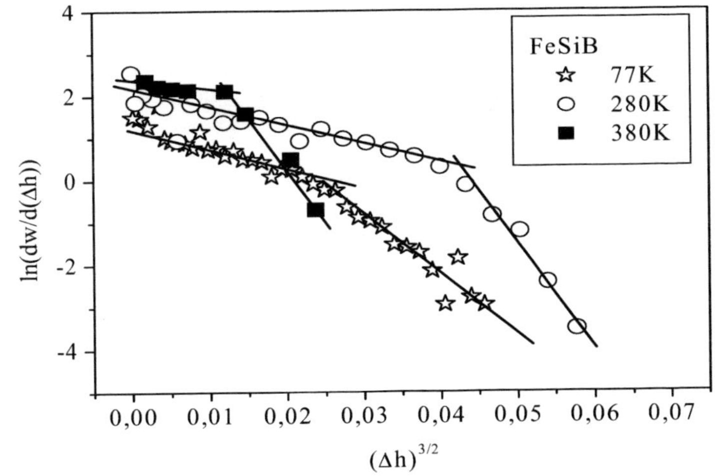

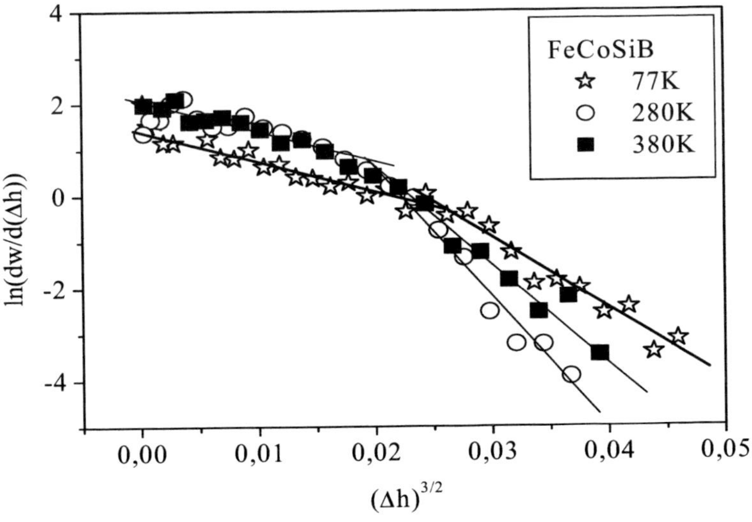

Figure 22. Continued on next page.

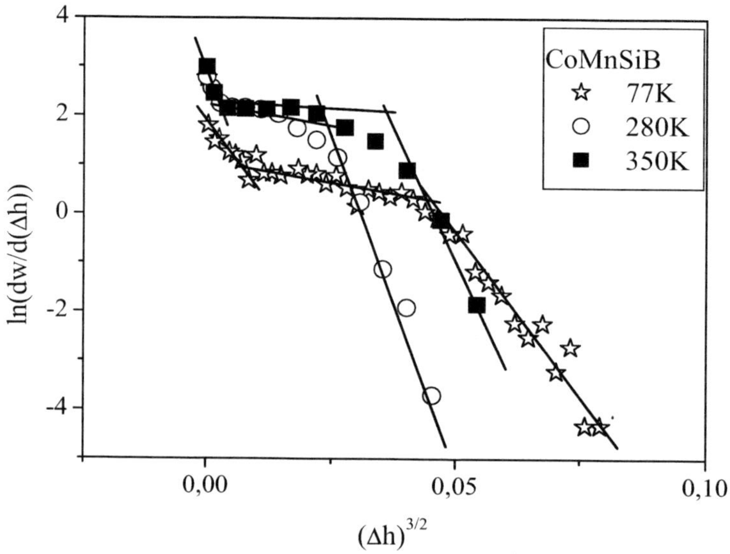

Reprinted with permission from [114], R. Varga et al., "Distribution and temperature dependence of switching field in bistable magnetic amorphous microwires" in Recent Res. Devel. Non-Crystalline Solids, 3 (2003) pp.85-95, ISBN: 81-7895-090-1, Publisher: Transworld Research Network, Trivandrum, India (2003) (figure 7).

Figure 22. Dependence of the logarithm of the probability density on the reduced field for FeCoSiB, CoMnSiB and FeSiB microwires.

Application of a mechanical stress moderates the magnetoelastic contribution to the domain wall potential. This results in the enhancement of the second local minimum in the domain wall free energy and even the decrease of the probability density with the increase of the external field has been found (figure 25).

Consequently, assumption on two contributions in the switchibng field distribution mechanism is reasonably explained by perfomed experiments

4.2.2. The Temperature Dependence of the Switching Field

From the temperature dependence of the position of the maximum in the switching field distribution we can estimate temperature dependence of the average switching filed and evaluate two contributions to the coercivity mechanism in the amorphous microwires.

The several mechanisms for magnetisation reversal through the domain wall displacements have been previously reported [117]. In particular, several sources of pinning mechanisms for the magnetisation process by wall displacements have been identified [118-119] contributing to the total coercivity. They are:

1. Volume pinning of the domain walls, DW, by structural defects, which is particularly important in magnetostrictive alloys,
2. Relaxation effects due to local structural rearrangements,
3. Clusters of chemically short-range ordered regions,

4. Surface irregularities
5. Intrinsic fluctuations of exchange energies and local anisotropy.

The effects are ordered in decreasing importance, the first one is assumed to have the biggest contribution to the coercivity.

The magnetoelastic contribution to the switching field $H_{sw}{}^{\sigma}$ can be expressed as [112-114]:

$$H_{sw}^{\sigma} \propto \frac{\sqrt{3A\lambda_s\sigma}}{M_s} \tag{37}$$

where M_s is the saturation magnetisation, A is exchange energy constant ($A \sim M_s^2$), λ_s is saturation magnetostriction, and σ is the total stress. In our case, $\sigma = \sigma_a(T) + \sigma_r$, where σ_r are residual intrinsic stresses coming from quenching and drawing, and $\sigma_a(T) \approx E(\alpha_g - \alpha_m)\Delta T$ are the stresses which presently depend on the temperature and arise from the different thermal expansion coefficients of the Pyrex, α_g, and metallic nucleus, α_m, and E is Young modulus. As early found in previous works [118], the temperature dependence of the magnetostriction of amorphous alloys can be described by the so-called scaling laws:

$$\lambda_s = k.M_s^{\varepsilon} \tag{38}$$

Here, the k is a constant, M_s is a saturation magnetisation and exponent ε depends on the origin of the magnetostriction. This exponent of the scaling law can show deeper information about the nature of the mechanism responsible for the magnetoelastic effects and the magnetic anisotropy [118, 120].

According to the [118-120], a M_s^3 dependence is expected for Fe- based amorphous alloys, when the magnetostriction originates from a single-ion anisotropy of uniaxial character. On the other hand, M_s^2 dependence can be observed in Co- based amorphous alloys, where the magnetostriction originates from an anisotropic two-ion exchange contribution of uniaxial symmetry. Finally in the samples with mixed composition (e.g CoFeSiB) the magnetostriction could be fitted by the linear combination of the given power laws: $\lambda_s = \alpha M_s^3 + \beta M_s^2$, where α, β are the constant coefficient, which meaning can be obtained elsewhere [121-124].

Thus, eq.(37) can be simplified into the form:

$$H_{sw}{}^{\sigma} \propto pM_s^x(1+r(\Delta T))^{1/2} \tag{39}$$

where $p \approx const.\sigma_r$, $r \approx E(\alpha_g - \alpha_m)/\sigma_r$, and x ∈ <1;3/2> - depends on the origin of magnetostriction mechanism.

It was also shown previously, that magnetoelastic pinning is not enough to fit the coercivity in the whole temperature range [125,126].

The amorphous alloys are produced by rapid quenching. As a result of their production their structure is in metastable state and for this reason quite large relaxation effects could be expected. Due to this relaxation phenomenon, an additional magnetic field is required in order to push the DW out of its position. This additional pinning field may contribute significantly to the total coercive field [118].

$$H_{sw}^{P}(T) \propto \frac{1}{M_s} \frac{\varepsilon_p^2 \rho_p}{kT} G(T,t), \tag{40}$$

where ε_p corresponds to the interaction energy of the mobile defects with the spontaneous magnetisation, ρ_p is the density of the mobile defects and $G(T,t)$ is a relaxation function [127]: $G(T,t) = (1 - e^{(-t/\tau)})$, where t is the time of measurement and τ is the relaxation time, which is given by the Arrhenious equation: $\tau = \tau_0 e^{Q/kT}$. Here, τ_0 is a pre-exponential factor given elsewhere [128] and Q corresponds to the activation energy of the mobile defects.

Two main contributions to the coercivity of amorphous alloys and its temperature dependence are then consider, the first one coming from the magnetoelastic interaction of the domain wall with the stresses and the second one coming from the magnetic relaxation due to the local structural rearrangement.

So, the general expression for switching fields in amorphous microwires takes the form of the sum of the above given contributions

$$H_{sw}(T) = H_{sw}^{\sigma} + H_{sw}^{P}(T). \tag{41}$$

Considering eq. (39) and (40) and applying those in (41) leads to:

$$H_{sw}(T) = pM_s^x(1+r(\Delta T))^{1/2} + n\, G(T,t)/(M_s T). \tag{42}$$

where $n \approx (\varepsilon_p^2 \rho_p/k)$ and is proportional to the number of the defects when $G(T,t)$ is assumed to be a monotonously increasing function of the measuring temperature at the measured times and temperatures [129].

Figure 26 shows the results of the fits according to the eq.42. We can see the effect of the different contributions according to the eq.39 and 40. The switching field distribution of the FeSiB microwire has strong increase of the coercivity at low temperature, which can be ascribed to the increase of the relaxation time of the mobile atomic defects. They loosing their mobility and acts as a pinning centers for the domain wall. With increasing the temperature, the mobility of the defects increases and the coercivity decreases. Above 170K, the magnetoelastic pinning dominates in the coercivity mechanism.

The different situation is in the FeCoSiB microwire, which has the highest magnetostriction. The magnetoelastic pinning dominates over all temperature range. The contribution from pinning on the atomic scale defects is ten times smaller then magnetoelastic one and can be registered only at lowest temperatures as a small increase of coercivity.

In the case of the CoMnSiB microwires, which have the smallest magnetostriction from the observed microwires, the pinning on the atomic scale defects also dominates at low temperatures. At higher temperatures the magnetoelastic pinning is prevailing, but the pinning on atomic scale defects is still comparable.

Consequently, two contributions to the coercivity mechanism in amorphous microwires have been suggested: the magnetoelastic one and stabilization one, coming from the magnetic relaxation. The two contributions have different temperature dependence which can be distinguished using the coercivity model.

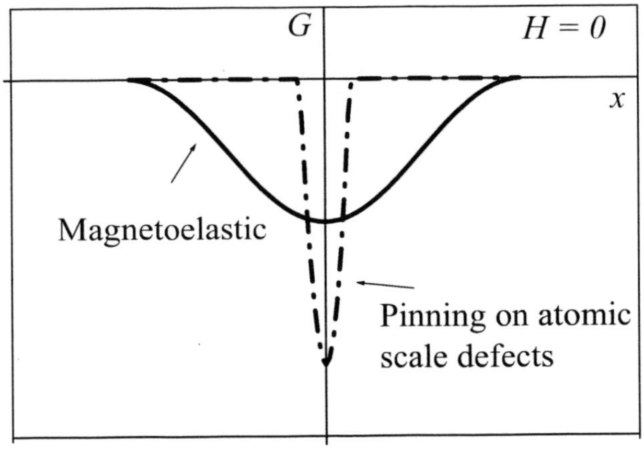

(a)

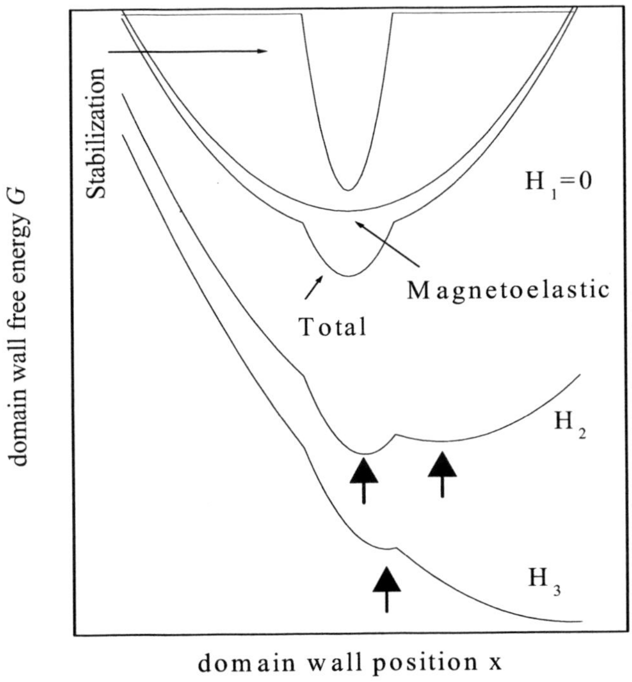

(b)

(a): Reprinted with permission from Ref. [114], R. Varga et al., "Switching-field distribution in amorphous magnetic bistable microwires" Phys. Rew. B 70, 024402 (2004), Copyright ©(2004) The American Physical Society (fig5); (b): Reprinted with permission from Ref. [116], R. Varga et al., "Magnetization processes in thin magnetic wires" J. Magn and Magn, Mater., 300, pp. e305-e310, (2006). Copyright Elsevier (2006). (figure 6).

Figure 23. Two potentials responsible for coercivity mechanism in magnetic microwire (a) and dependence of the free energy of the closure domain wall on its position (under the action of external field fields $H_1<H_2<H_3$). The arrows point to the local minima.

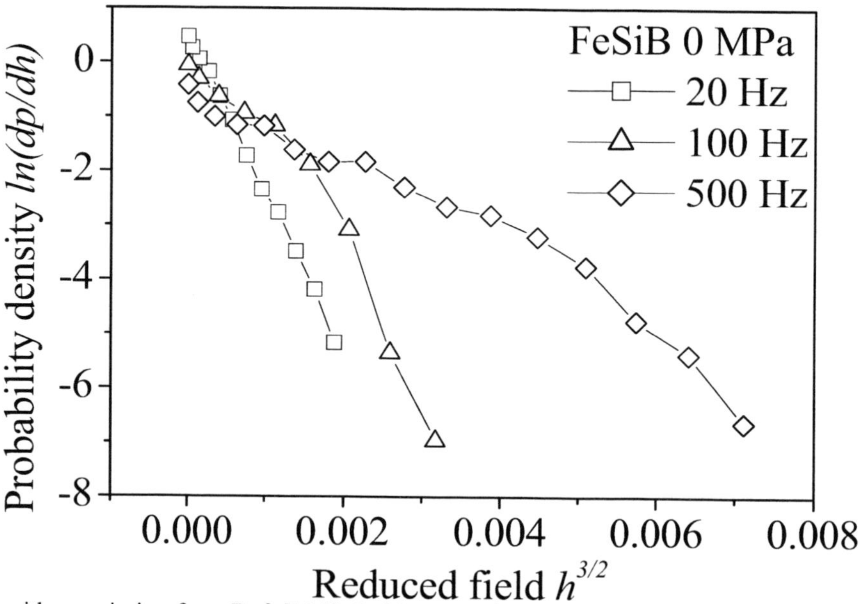

Reprinted with permission from Ref. [116], R. Varga et al., "Magnetization processes in thin magnetic wires" J. Magn. Magn. Mater., 300, pp. e305-e310, (2006). Copyright Elsevier (2006) (figure 7).

Figure 24. Renormalized switching field distribution of FeSiB microwire measured at different frequencies.

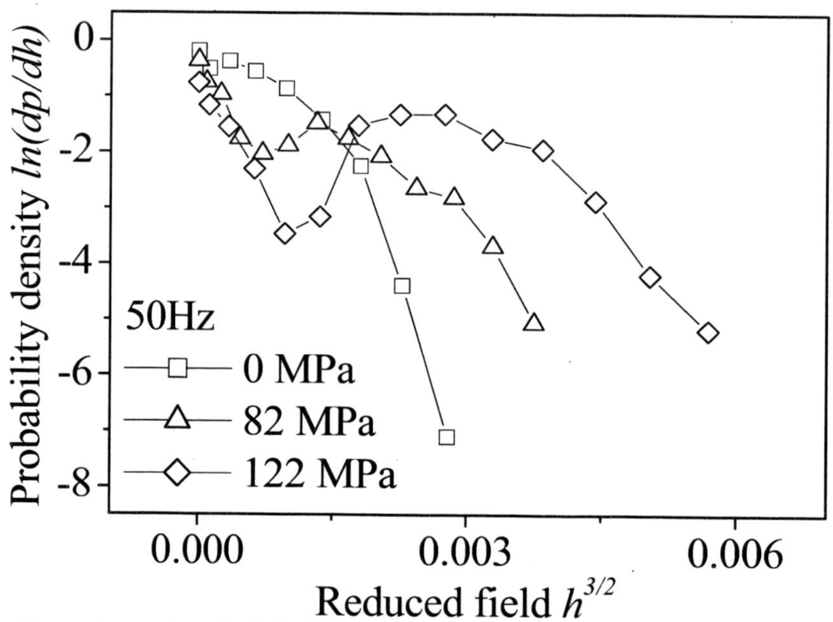

Reprinted with permission from Ref., [116], R. Varga et al., "Magnetization processes in thin magnetic wires" J. Magn. Magn. Mater., 300, pp. e305-e310, (2006). Copyright Elsevier (2006) (figure 8).

Figure 25. Renormalized switching field distribution of FeSiB microwire measured under different applied load.

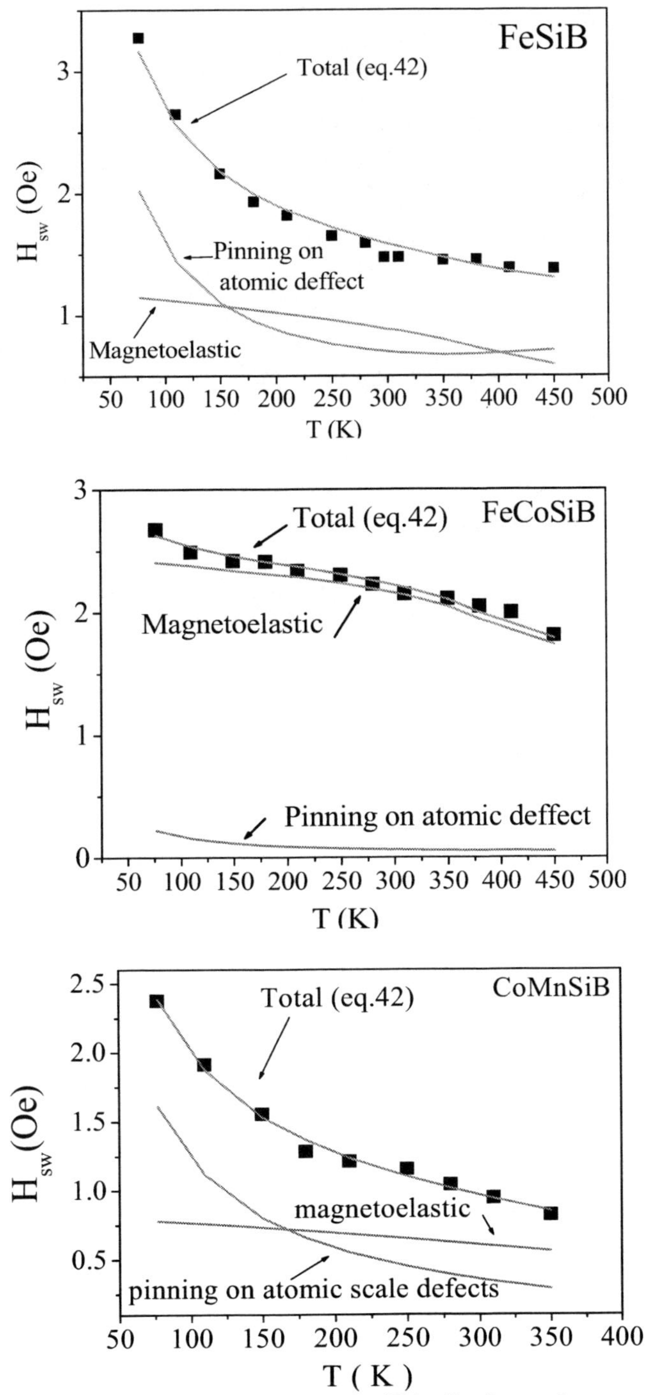

Reprinted with permission from [114], R. Varga et al., "Distribution and temperature dependence of switching field in bistable magnetic amorphous microwires" in Recent Res. Devel. Non-Crystalline Solids, 3 (2003) pp. 85-95, ISBN: 81-7895-090-1, Publisher: Transworld Research Network, Trivandrum, India (2003) (figure 5).

Figure 26. Temperature dependence of the mean switching field for microwires with different magnetostriction.

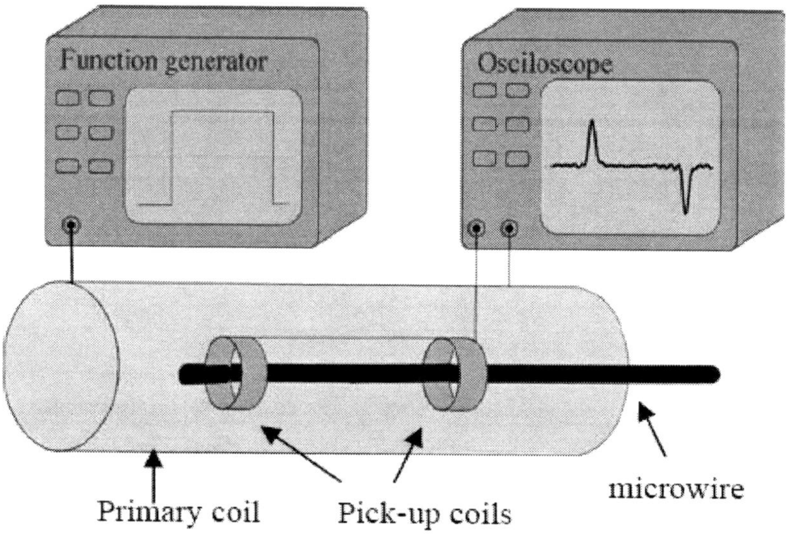

Reprinted with permission from Ref. [15], R. Varga et al. "Fast magnetic domain wall in magnetic microwires" Phys. B 74, 212405 (2006). Copyright American Physical Society (2006) (figure 2).

Figure 27. Schematic representation of the experimental set-up for measuring DW velocity.

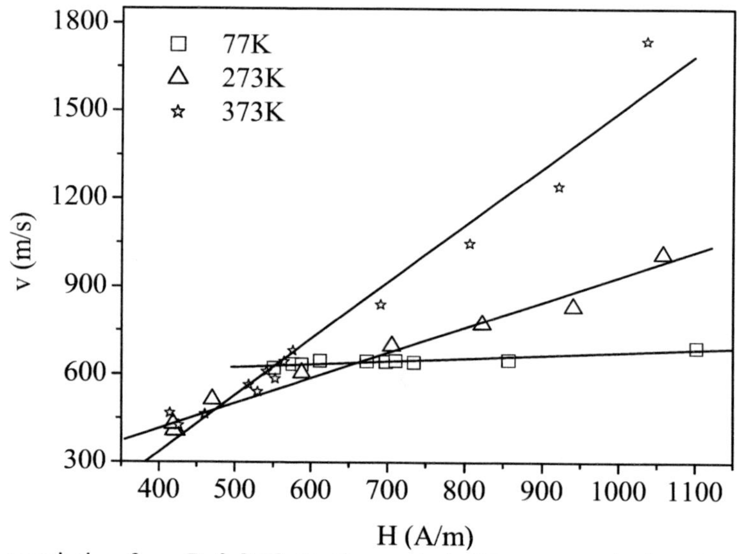

Reprinted with permission from Ref. [15], R. Varga et al. "Fast magnetic domain wall in magnetic microwires" Phys. Rev. B 74, 212405 (2006). Copyright American Physical Society (2006) (figure 3).

Figure 28. Domain wall velocity of $Co_{40}Fe_{36}Si_{1.1}B_{12.9}$ microwires as a function of magnetic field amplitude measured at different temperatures.

From the dependence of the logarithm of the probability density to switching on reduced magnetic field we have also found two contributions to the domain wall pinning. These were attributed to the magnetoelastic and stabilisation interaction.

4.2.3. Domain Walls Propagation

As pointed above, one of the characteristic features of the magnetic bistability is the appearance of rectangular hysteresis loop related with single and large Barkhausen jump appeared at low applied magnetic field. Such magnetic bistable behavior has been related to the presence of a single Large Barkhausen Jump, which was interpreted as the magnetization reversal in a large single domain [13, 49, 99, 106, 108-110]. The rectangular hysteresis loop could be interpreted in terms of nucleation or depinning of the reversed domains inside the internal single domain and the consequent domain wall propagation [48, 53, 108-114, 130]. Perfectly rectangular shape of the hysteresis loop has been related with a very high velocity of such domain wall propagation.

Recently great attention has been paid to studies of domain wall (DW) propagation in different materials, among them in thin wires with sub-micrometric and micrometric diameter [131-148]. In fact, domain wall propagation has been observed in different kinds of magnetic materials with rectangular hysteresis loop. Particularly, domain wall velocity has been successfully measured in conventional amorphous wires and Ni single crystals [132,133]. Recent growing interest on DW propagation is related with proposals for prospective logic [131] and memory devices [134]. In these devices, information is encoded in the magnetic states of domains in lithographically patterned nanowires. DW motion along the wires allows for the access and manipulation of the stored information. The speed at which a DW can travel in a wire has an impact on the viability of many proposed technological applications in sensing, storage, and logic operation [134]. When a DW is driven by a magnetic field parallel to the long axis of the wire, the maximum wall speed is found to be function of, and capped by, the dimensions of the wire [131,135]. In addition, the range of applied driving fields for which the wall motion is pure, fast translation is also limited by the wire geometry. Special effort has been performed in nanowires to enhance the DW speed. Thus, application of transverse magnetic field proposed in [135] allowed to improve slightly the DW speed till about 600 m/s at about 120 A/m.

One of the most interesting properties of glass-coated microwires is that related with the observation of spontaneous magnetic bistability in Fe-rich compositions [2-5]. This rectangular hysteresis loop was interpreted in terms of nucleation or depinning of the reversed domains inside the internal single domain and the consequent domain wall propagation [14, 136,137]. Additionally, it was found that small closure domain appears at the end of the wire, in order to decrease the magnetostatic energy [13,53]. Such peculiar domain structure appears because of the stresses originating from the simultaneous quenching of the metallic nucleus surrounded by the glass coating with the different thermal expansion coefficient of the glass coating and metallic nucleus.

Quite high DW velocities achieving up to 18 km/s have been recently reported for glass-coated microwires with few μm diameter [6]. Perfectly rectangular shape of the hysteresis loop has been related with a very high speed of such domain wall propagation. It is demonstrated by few methods that the remagnetization process of such magnetic microwire starts from the sample ends as a consequence of the depinning of the domain walls and subsequent DW propagation from the closure domains [13, 136,137]. Among the most unusual results observed in these microwires, the DW propagation below the switching field [14] and negative critical propagation field [15] should be mentioned.

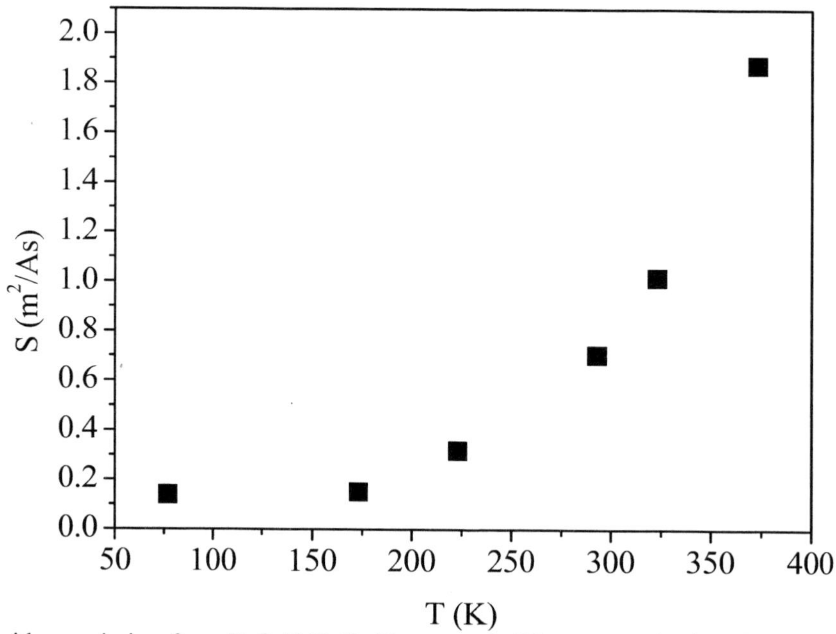

Reprinted with permission from Ref. [15], R. Varga et al. "Fast magnetic domain wall in magnetic microwires" Phys. Rev. B 74, 212405 (2006). Copyright American Physical Society (2006). (figure 4)

Figure 29. Temperature dependence of the domain wall mobility of $Co_{40}Fe_{36}Si_{1.1}B_{12.9}$ microwires.

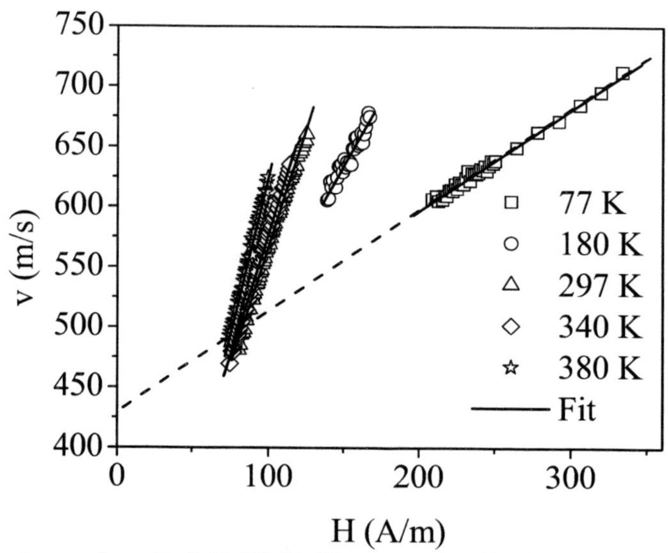

Reprinted with permission from Ref. [142], R. Varga et al. "Single-DomainWall Propagation and Damping Mechanism during Magnetic Switching of Bistable Amorphous Microwires". Phys. Rev. Lett. 94, 017201 (2005). Copyright (2005) The American Physical Society (figure 2).

Figure 30. Dependence of domain wall velocity on applied magnetic field for FeSiB microwires. The temperature as a parameter.

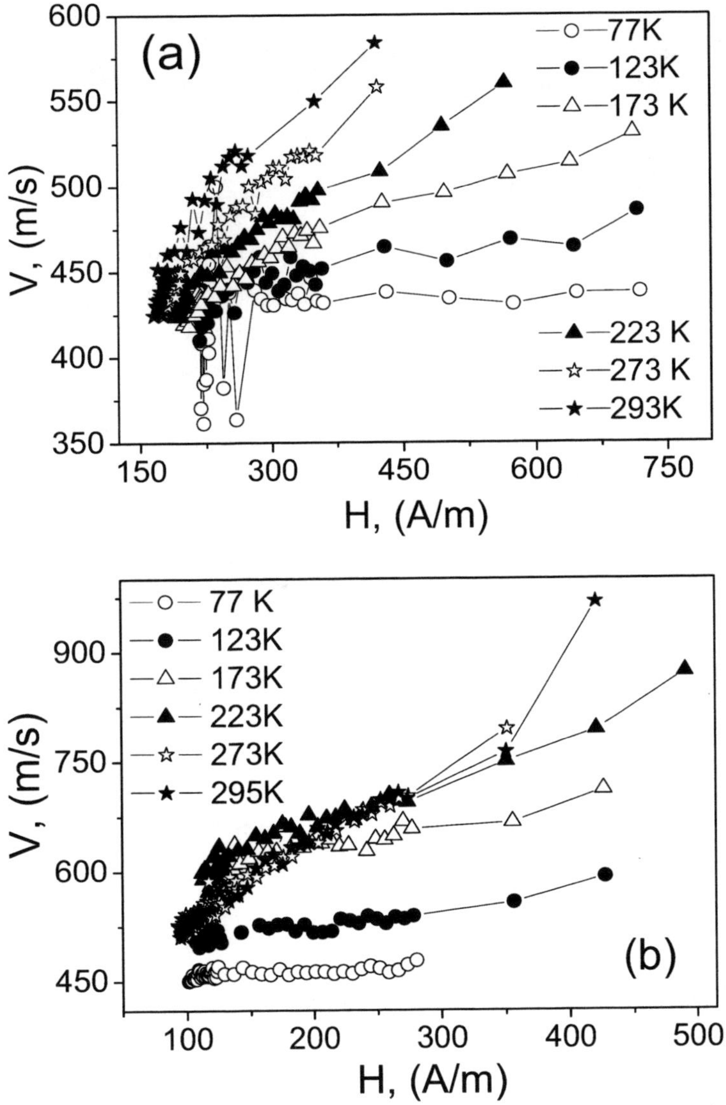

Reprinted with permission from Ref. [16] V. Zhukova et al. "Domain wall propagation in thin Fe-rich glass-coated amorphous wires" Phys. Status Solidi A *4*, p. 679-682 (2009). Copyright WILEY-VCH Verlag GmbH & Co. KGaA (2009). (figures 2, 4)

Figure 31. Magnetic field dependence of DW velocity in $Fe_{69}Si_{10}B_{15}C_6$ sample with d= 14 μm (a) and in $Fe_{69}Si_{10}B_{15}C_6$ sample with d= 18 μm measured at different temperatures with fixed frequency, f=*100 Hz*, of AC exiting magnetic field.

In the case of microwires the DW dynamics was measured by the classical Sixtus Tonks [138] experiments, as described recently elsewhere [6, 14, 15]. Usually the system consists of three coaxial coils (Figure 27): two detecting coils and one nucleation coils. Additional pick-up coils can be placed in order to study more detailed the domain wall propagation along the microwire. In some case nucleation coil did not used and one end of the sample has been placed outside the solenoid in order to ensure domain wall nucleation always near one of the microwire ends.

The domain wall dynamics in a viscous medium and at slow rates is governed by a mobility relation [138,139]:

$$v=S\ (H-H_0), \tag{43}$$

where S is the domain wall mobility and H_0 is the critical propagation field, below which the domain wall propagation is not possible.

Domain wall propagation has been studied in glass-coated microwires with positive magnetostriction constant with chemical composition and geometry (ratio, ρ, between metallic nucleus diameter, d, to total diameter D).

Figure 28 shows the domain wall velocity as a function of the applied magnetic field measured in $Co_{40}Fe_{36}Si_{1.1}B_{12.9}$ microwire. The domain wall velocity in $Co_{40}Fe_{36}Si_{1.1}B_{12.9}$ microwires is quite high reaching the values up to 1800m/s at 1050 A/m at the temperature 373K. A key dynamic parameter for the domain wall motion is the domain wall mobility *S,* e.g. the rate of change of wall velocity *v* with increasing external magnetic field *H.*[2] The average domain wall mobility (taken from the linear fit of the data in Figure 28) in $Co_{40}Fe_{36}Si_{1.1}B_{12.9}$ microwires varies with the temperature from 0.14 m^2/As at 77K up to 1.93 m^2/As at 373 K (Figure 29) in contrary to the domain wall mobility for tri-layer sub-micron magnetic wire [140].

The domain wall mobility measured at 373K is five times higher than the mobility measured in Ref. 141. On the other hand, the mobility at low temperature (77K) is lower. Anyway, the velocity remains quite high (over 600m/s) in whole range of the applied field.

Initially linear magnetic field dependence, v(H) has been observed at some magnetic field range [142]. Later it was shown that extending the magnetic field range, v(H) dependences are essentially non-linear [16].

Figure 30 shows linear dependence of the wall velocity on the applied magnetic field observed in $Fe_{77.5}Si_{7.5}B_{15}$ microwire. A number of considerations have been performed extrapolating the linear behavior (as indicated for the case of 77 K) to zero applied field and zero velocity, the velocity, v_0, of the wall at zero applied field and the field, H_0, at zero velocity has been estimated. Positive values of v_0 and negative values of H_0 (up to -500 A/m at 77K) were underlined. This result was interpreted considering even that the domain wall can theoretically move even in negative applied field, or that negative loses could exist during the wall propagation. Existence of reverse domains at negative fields at the ends of the microwire has been also considered.

Later, essentially non-linear v(H) dependences have been observed in a number of microwires compositions.

The dependence of *v* on applied magnetic field, *H,* measured for the $Fe_{69}Si_{10}B_{15}C_6$ microwire with the diameter of metallic nucleus 14 μm at different temperatures, *T,* is shown in Figure 31a. It is well seen, that *v(H)* dependence is essentially not linear, at all temperatures, showing significantly higher DW mobility, *S=dv/dH,* at low filed limit. Such non-linearity of *v(H)* is most significant at lower temperatures. Roughly, *v(H)* dependence of this microwire ($Fe_{69}Si_{10}B_{15}C_6$ with *d*=14 μm and *D*=33 μm) can be described as consisting of 2 linear *v(H)* dependences, increasing DW mobility, *S,* for both regions with increasing the temperature.

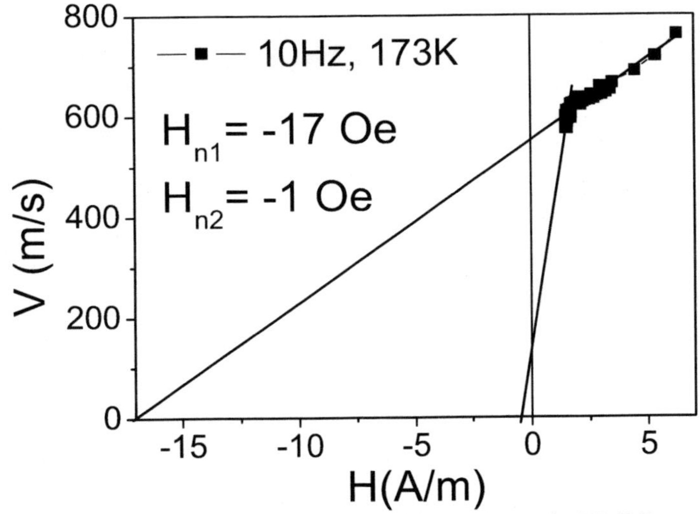

Figure 32. Magnetic field dependence of domain wall velocity in $Fe_{69}Si_{10}B_{15}C_6$ sample with d=18 μm with approximation to *v*=0 for low and high field regimes.

Similarly, *v(H)* dependences measured at the same conditions of experiment for $Fe_{69}Si_{10}B_{15}C_6$ microwire with *d*=18 μm and *D*=23.4 μm in temperature range 77-295K presented essentially non-linear dependences for all the temperatures (Figure 31b).

It is worth mentioning that such non-linear shape of *v(H)* dependences observed for both microwires introduced some corrections in the estimations of the critical propagation field recently reported elsewhere and giving large negative values. Indeed the approximation to zero domain wall velocity from the region 2 at intermediate field (see Figure 32) gives large negative critical propagation field, $H_{n1}\approx$-17 Oe for the case of the sample with metallic nucleus 18 μm and the ratio ρ=0.77 measured at 173K. On the other hand, such approximation taken from the lowest field regime gives values in vicinity to zero magnetic field ($H_{n2}\approx$-1 Oe). As has been mentioned above, such non linearity observed at low field can be attributed to the change of the DW structure with changing of the applied magnetic field. In fact the estimation of the critical propagation field should be taken for the domain wall existing for the lower field. As has been mention above, such estimation gives the values H_n about 0.

Observed non-linearity of *v*(*H*) dependences related with change of the regime from the lower field with smaller DW velocity to the higher field with enhanced velocity at some critical field, H_{cr}, can be attributed to the change of the structure of the DW as well as to the so-called adiabatic domain wall propagation at magnetic field slightly above critical field [143]. Such adiabatic regime was attributed to the interaction of the DW with the local defects inside the sample. The existence of such local defect has been recently confirmed when fluctuations of the local nucleation fields has been observed [149]. Similar explanation has been recently considered in ref. [144].

Regarding the change of the DW mobility and/or domain wall velocity due to the change of the DW structure, it was shown previously by micromagnetic simulations for the case of nanowires, that the vortex-type domain wall is faster than transversal one [145]. Additionally, the vortex domain wall is more stable for the case of thicker magnetic wires than the transversal one [146].

On the other hand consideration of the so-called adiabatic domain wall propagation at magnetic field slightly above critical field led to assumption that the critical field H_0 consists of two parts: long range contribution H_l arising from the magnetostatic and magnetoelastic interaction of the domain wall with the stresses. On the other hand, the domain wall interacts during its propagation with the local defects giving rise to the short range contribution to the critical field H_p, which corresponds to the pinning field of the domain wall on the defects distribution. Finally, the domain wall dynamics is described by [143]:

$$v = S(H-(H_l+H_p)) \qquad (44)$$

As a result, the domain wall moves in the small intermittent jumps at a velocity, which is given by the local distribution of the defects according to eq.44. Such a regime of the domain wall movement is called adiabatic.

The distribution of the pinning field H_p can be approximated by the gaussian shape with the width of R [147]. According to the Random-Field Ising model, the magnetization change ΔM during the small domain jump is given by the power law [147]:

$$\Delta M \sim (R-R_c/R_c)^{\beta} \qquad (45)$$

where R_c is the critical distribution width, below which the small intermittent domain wall jumps do not appear and β is the critical exponent. The average velocity of the domain wall finally scales [148]:

$$v=S'(H-H'_0)^{\beta} \qquad (46)$$

where S' is the so-called domain wall mobility parameter and H'_0 is the critical field, below which the domain wall movement cannot be observed.

It is worth mentioning that usually the magnetization switching in magnetically bistable microwires is associated with single domain wall propagation, although the micromagnetic origin of this remagnetization front is still unclear.

As it it stated in a number of published papers, for applications of domain wall propagation in magnetic memories it is essentially important not only fast domain wall propagation itself, but also interaction of domain wall with defects, i.e. controllable domain wall pinning on defects. Therefore from the point of view of future applications it is quite important to understand the role of defects in domain wall propagation.

Recently local domain wall nucleation using short nucleation coil has been realized in glass-coated microwires [149]. It was shown particularly that this local nucleation field, H_N, far from the wire ends is rather higher than the switching field observed in bulk hysteresis loop (3-4 times higher than bulk switching field). This is because usually domain wall propagation starts from the closure domains arising at the wire ends in order to diminish the stray fields.

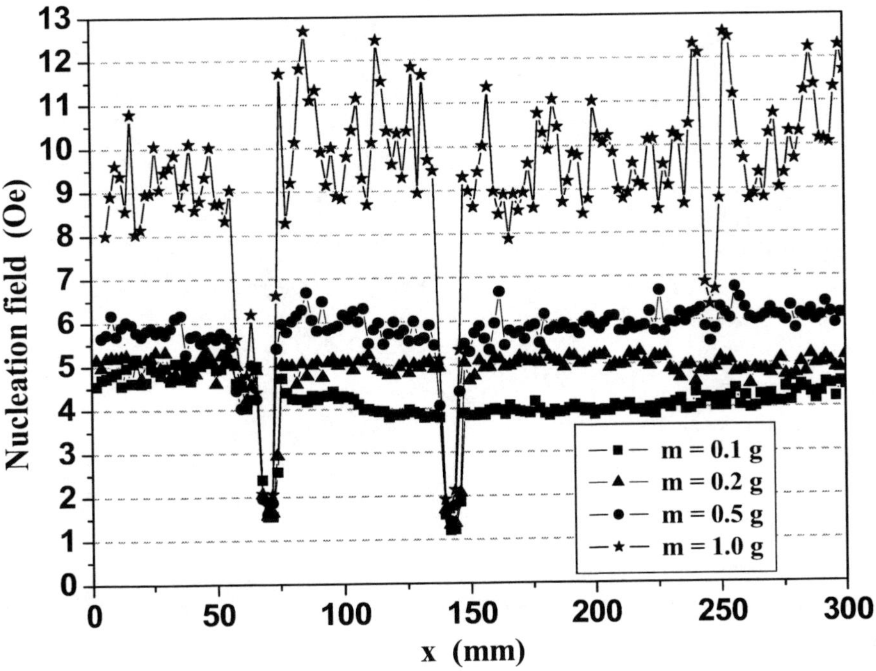

Reprinted with permission from Ref. [149] M. Ipatov et al. "Local nucleation fields of Fe-rich microwires and their dependence on applied stresses" Phys. B. 403, 379-381, 2008. Copyright Elsevier (2008) (figure 1).

Figure 33. Distribution of the local nucleation fields along the length of $Fe_{74}Si_{11}B_{13}C_2$ wire with radius $r_0 = 6$ μm and $\rho = 0.76$ for different mass loads.

Figure 33 shows the local nucleation fields of $Fe_{74}Si_{11}B_{13}C_2$ microwire as function of the wire length for different mass loads attached to the wire. The deep holes in the curves presented in Figure 33 correspond to the positions of strong localized defects existing within the wire. In fact this behaviour is typical for all studied microwires with different compositions. The present case is typical and even measurements have been performed in one of the most homogeneous microwires. For the particular case of domain wall propagation, quite recently it was shown using few pick-up coils that such defects can serve as the domain wall nucleation centers, when the applied magnetic field is high enough, i.e. comparable with the lowest local nucleation field within given peace of microwire [150]. Below lowest nucleation field roughly linear v(H) has been found. Therefore, deviations from linearity v(H) can be also related with multiple domain wall nucleation.

Consequently, it is worth distinguishing an interesting mechanism of magnetization reversal in magnetically bistable microwires. This mechanism is conditioned by the microwires inhomogeneities. These inhomogeneities can lead to considerable acceleration of the sample magnetization switching. On the other hand, neglecting of this factor can result in exaggerated estimation of DW velocity from Sixtus-Tonks experiment. We believe that at magnetic field above H_N contribution of defects can be essential. In this case under the action of external magnetic field, a new reversed domain can be spontaneously nucleated in front of the propagating head-to head closure domain. Appearance of these additional domains at $H > H_N$ can accelerate remagnetization switching.

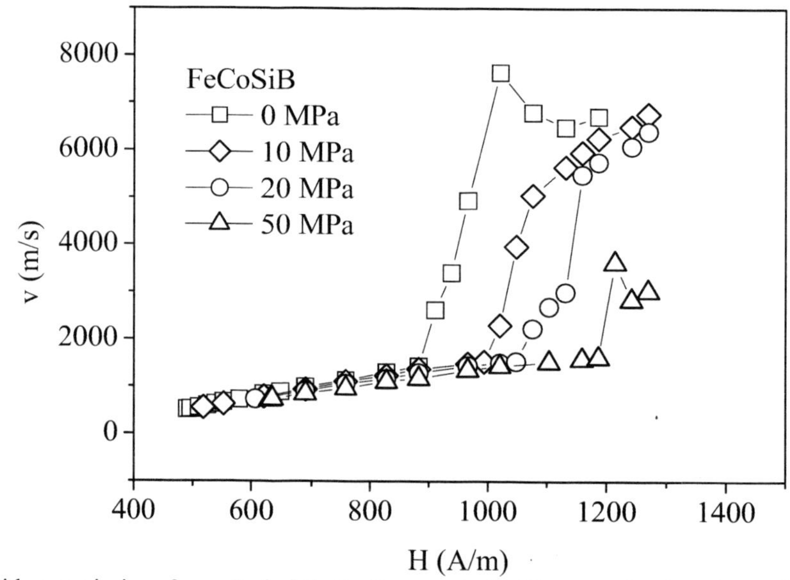

Reprinted with permission from Ref. [6] R. Varga et al. "Supersonic domain wall in magnetic microwires" Phys.Rev.B. V 76, 132406, 2007. Copyright ©2007 The American Physical Society (figure 2).

Figure 34. Domain wall velocity v as a function of the magnetic field amplitude H for FeCoSiB microwire with the applied tensile stress as a parameter.

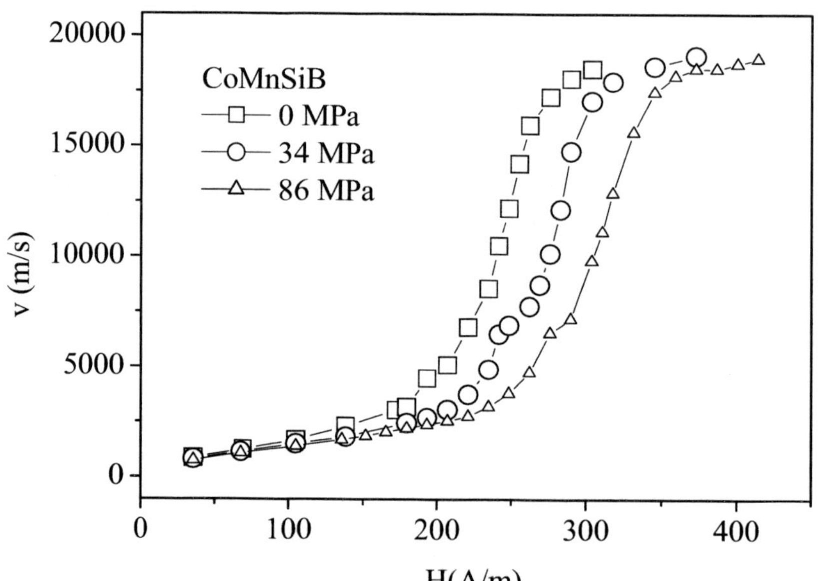

Reprinted with permission from Ref. [6] R. Varga et al. "Supersonic domain wall in magnetic microwires" Phys.Rev.B. V 76, 132406, 2007. Copyright ©2007 The American Physical Society (figure 3).

Figure 35. Domain wall velocity v as a function of the magnetic field amplitude H for CoMnSiB microwire with the applied tensile stress as a parameter.

Observed stress dependence of local nucleation field also allows estimation of the magnetostriction constant, λ_s , [149] using relation of local nucleation field, H_n and , λ_s:

$$H_n = -\frac{3\lambda_s}{M_s}\left(\sigma_{zz} - \sigma_{\varphi\varphi} + \sigma_{app}\right), \tag{47}$$

where σ_{zz} - $\sigma_{\varphi\varphi}$ is the difference of the longitudinal and circumferential residual stress tensor components of the amorphous wire, σ_{app} is the amplitude of the applied tension stress.

In any case non-linearity observed at high field regimes still needs further interpretation. On the other hand strong interest in DW propagation resulted in detailed and extended studies of DW propagation in microwires. Thus, quite fast domain wall propagation has been reported in ref. [5] for $Fe_{36}Co_{40}Si_{11.1}B_{12.9}$ and $Co_{68}Mn_7Si_{10}B_{15}$ microwires three regions of the domain wall dynamics were clearly recognized in $Fe_{36}Co_{40}Si_{11.1}B_{12.9}$ microwire (Figure 34). Firstly, the linear dependence of the domain wall velocity was found at low fields (0-900 A/m) with the domain wall mobility of 2.3 m^2/A.s up to the velocity of 1400 m/s. In the second range (900-1000A/m), the domain wall mobility increases 20 times, reaching the value of 45 m^2/A. The domain wall velocity increases steeply in this range reaching its maximum at 7700 m/s. Finally, the domain wall mobility becomes negative and the domain wall velocity decreases. The most probable explanation for the slope change has been attributed to the change of the domain wall structure. Few mechanisms, such as existence of transverse anisotropy, low eddy current damping have been discussed in order to explain high DW velocity observed in glass-coated microwires.

The domain wall dynamics becomes more interesting in the case of amorphous $Fe_{49.6}Ni_{27.9}Si_{7.5}B_{15}$ microwire (Figure 35). At the field of around 600 A/m, the domain wall velocity and mobility steeply increase [151].

Observed non-monotonity on V(H) dependence (Figure 35) have been attributed in [151] to the interaction of DW with the sound, when the domain wall approaches the sound speed limit. Since the amorphous bistable microwires are highly (with respect to soft magnetic materials) magnetostrictive, the domain wall produces elastic wave when it propagates. Such a wave always accompanies the propagating domain wall in magnetostrictive materials. Elastic waves in magnetic microwires are very fast (The speed of elastic wave in amorphous microwires has been found to be around 4700 m/s [152,153]) and they disappears much faster then domain wall propagation. However, when the domain wall approaches the sound speed limit, the dissipation in the elastic subsystem of a microwire grows substantially and the DW transfers part of its magnetic energy to the elastic subsystem [154]. As a result, the domain wall velocity remains constant. The magnetic energy E transferred to the elastic subsystem can be calculated from the constant part of the domain wall velocity dependence $v(H)$ [155]:

$$E = 2\mu_0 M_s \Delta H, \tag{48}$$

where ΔH is the width of the constant part in $v(H)$ dependence. The transferred magnetic energy increases the amplitude of the elastic waves until the elastic waves damping becomes equal to $2\mu_0 M_s \Delta H$. Finally, the domain wall exceeded the sound speed when the driving field becomes greater than the force of dynamic breaking of the domain wall by the elastic waves. Then, domain wall velocity steeply jumps to higher values.

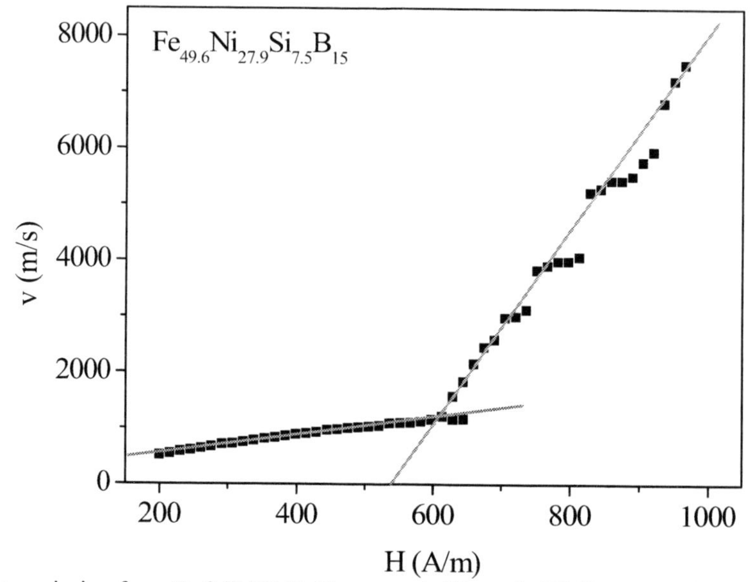

Reprinted with permission from Ref. [151] R. Varga et al. "Domain Wall Propagation in Thin Magnetic Wires" IEEE Trans. Magn., V 44, N° 11, pp. 3925-3930, 2008. Copyright IEEE Magnetic Society (2008) (figure 5).

Figure 36. Dependence of the domain wall velocity v in amorphous $Fe_{49.6}Ni_{27.9}Si_{7.5}B_{15}$ microwire on the applied magnetic field H. Full line corresponds to the fit according to eq.43.

Such effect is already well known in ferrites [154-156] that have much smaller magnetization comparing to the amorphous Fe-based microwires. However, this is the first time when supersonic domain wall are observed in ferromagnetic wires.

Generally, the sound velocity s in solids is given by:

$$s = (E/\omega)^{1/2}, \qquad (49)$$

where E is Young modulus and ω is mass density. In a solid, there is a non-zero stiffness both for volumetric and shear deformations. Hence, it is possible to generate sound waves with different velocities dependent on the deformation mode.

In amorphous $Fe_{49.6}Ni_{27.9}Si_{7.5}B_{15}$ microwire, the interaction of the domain wall with the elastic waves appears three times at the velocities of 3130, 4070 and 5430 m/s respectively (figure 36). These velocities correspond well to the sound velocitiy obtained for Fe-based microwires (4700 m/s, [154-156]) or to the sound velocity for Crystalline Iron (5100 m/s). The existence of three sound velocities can be attributed to different sound waves that can be created in a finite solid body. For example, three phonon waves were measured in stainless steel rod: longitudinal, transversal and extensional one. The values measured for microwires corresponds almost perfectly (±5%) to the longitudinal (5790 m/s) and transversal (3100m/s) phonon wave. However, sound speed (4070 m/s) either does not represent the extensional wave, or one should take into account the stress applied by the glass coating as well as different dimensions of the microwires in order to explain discrepancies.

The maximum domain wall velocity is fixed by the restoring forces that oppose precession inside the DW (Walker model) [157]. Thus, it could be increased by adding an anisotropy with transverse easy axis. Theoretically, it was shown, that the domain wall velocity increases with the presence of the transversal anisotropies [158, 159]. Glass-coated microwires exhibit strong radial anisotropy arising from the stresses induced by the glass coating through the difference in the thermal expansion coefficients of the metallic nucleus and the coated glass. In the case, when both anisotropies (axial as well as transversal) are compensated, the restoring force that opposes precession inside the DW is minimum. This happens just at the boundary in between the axial and radial domain structure. Additionally, the domain wall does not interact with the surface of the wire (due to the presence of the radial domain structure) avoiding the surface pinning of the domain wall. The ends of the propagating domain wall are moving at the interface between the axial and radial domain structure, where both axial and radial anisotropies are exactly compensated. This surely helps for the domain wall to achieve the fast propagation.

As a result of their amorphous nature, the glass-coated microwires have high resistivities. Therefore, the eddy current damping is very low in these materials [119]. The most important damping mechanism in glass-coated microwires at room temperature was found to arise from spin relaxation. It was shown in ref. 132, that damping contribution coming from the spin relaxation (β_r) is proportional to the square root of the anisotropy [132]. In the case of amorphous glass coated microwire the anisotropy is determined mainly by the magnetoelastic interaction of the local magnetic moments with the stresses σ introduced during the microwire's preparation (so, $\beta_r \sim (\lambda_s \sigma)^{1/2}$, where λ_s is the saturation magnetostriction).

Even higher DW propagation has been observed in $Co_{68}Mn_7Si_{10}B_{15}$ (with the diameter of the metallic core being 8 μm and total diameter being 20 μm) which is characterized by the lowest possible magnetoelastic anisotropy (due to its low magnetostriction λ_s) among the microwires with single domain axial structure (see Figure 36). In fact it has 10 times smaller magnetoelastic anisotropy (in comparison to the FeCoSiB microwire) and therefore extremely small domain wall damping. Anyway, the stresses applied on the metallic nucleus are of the same directions like in the FeCoSiB microwire and it presents the same domain structure.

As a result of the small damping, the domain wall mobility is extremely high (16 m^2/A.s at low fields) and it reaches the maximum value of 272 m^2/A.s at the field of 255 A/m (Figure 35). Although the Walker field is small (~300A/m), the maximum velocity reaches a value of 18 500 m/s. It was assumed, that the maximum velocity is given not only by the transversal anisotropy but mainly by the high mobility as well as by the counterbalance between the uniaxial and transversal anisotropy [158,159].

Application of a tensile stress shifts the Walker fields to higher values, but the effect on the maximum velocity is not so strong as in the case of FeCoSiB microwire. We think that this behaviour is as a consequence of the magnetoelastic anisotropy ($E_\sigma \sim \lambda_s \sigma$) depending on the magnetostriction, which is quite small [119]; in this case $\lambda_s \sim 10^{-7}$.

Summarizing, quite fast domain wall propagation has been observed in glass-coated microwires. Essentially non-linear dependence of the DW speed on magnetic field has been observed. The origin of such dependences and high domain wall velocity is discussed in terms of magnetoelastic energy contribution (particularly of transverse magnetic anisotropy contribution) and interaction of domain walls with internal defects as well as adiabatic domain wall propagation. Certainly, fast domain wall propagation found in glass-coated microwires is quite interesting for technological applications. Possible role of defects is related with nucleation centers of new reversed domains.

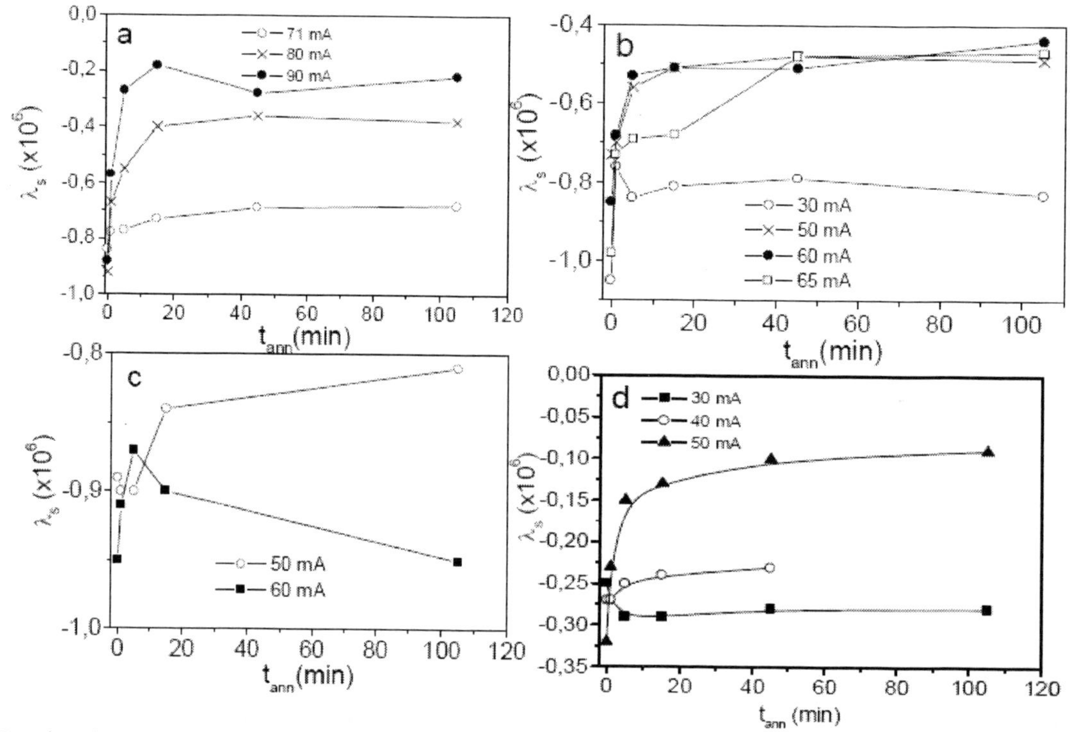

Reprinted with permission from Ref. [116], A. Zhukov at al., "Magnetostriction in glass-coated magnetic microwires" J. Magn and Magn, Mater., 258-259, pp.151-157 (2003). Copyright Elsevier (2002) (figure 3).

Figure 37. Dependences of magnetostriction constant, λ_s, on annealing time, t_{ann} , for glass-coated amorphous $Co_{57}Fe_{6.1}Ni_{10}B_{15.9}Si_{11}$ (a), $Co_{67.5}Fe_4Ni_{1.5}B_{14}Si_{12}Mo_1$ (b), $Co_{69.1}Fe_{5.2}Ni_1B_{14.8}Si_{9.9}$ (c) and $Co_{69.5}Fe_{3.9}Ni_1B_{12.8}Si_{10.8}Mo_2$ (d) microwires.

Switching field fluctuations have been observed and explained using thermo-activation approach and considering the magnetoelastic and stabilisation potentials as was also deduced from the temperature dependence of the coercivity. The magnetoelastic potential comes from the long range interaction. On the other hand, the potential coming from pinning on atomic scale defects has short-range origin.

4.3. Giant Magneto-impedance Effect and Enhanced Magnetic Softness. Tailoring of Magnetic Properties and GMI

During the last few years the giant magneto-impedance effect, GMI, became a topic of great interest in the field of applied magnetism owing to the large sensitivity of the total impedance with the applied *DC* field at low magnetic fields and high frequencies [1-4, 7, 9, 13]. This GMI effect consists in a large change of the electric impedance of a magnetic conductor when is subjected to an axial *dc* magnetic field, H. It has been recognized that the large sensitivity of the total impedance of a soft magnetic conductor at low magnetic fields and high frequencies of the driven *ac* current originates from the dependence of the transverse magnetic permeability upon the *dc* magnetic field and skin effect. Such sensitivity achieves up to 300% relative change of the impedance in conventional amorphous wires with vanishing magnetostriction. Recently large GMI effect (up to 600%) has been also found in glass-coated microwires [91, 136, 160-163]

Since the re-discovery of the magneto-impedance effect in 1994, a number of researches and developments in this direction have been widely performed by different groups. Thus, a CMOC IC circuitry with pulse current operation was established for the MI sensors [7,8]. The stress-impedance (SI) and torsion impedance (TI) effects showing high sensitivity of the impedance to the applied stress with a strange gauge factor of 2000-4000 has been found in amorphous wires [7,8, 164-170].

As a result, the GMI and SI sensors with the CMOC IC circuitry with advantageous features comparing with conventional magnetic sensors have been developed by different companies [2,7,8]. Main applications are related with the detection of the magnetic fields, small weights and vibrations, and such branches of the industry as the car industry and medicine are main consumers of these sensors. Among the almost 100 applications, the following sensors are currently in progress: portable digital display of the terrestrial magnetic field, brain tumor sensor, secondary current sensor for the induction motor control, car passing measurement and recording disk, finger-tip blood vessel pulsation, mechanoencephalogram sensor, etc

Such enhanced GMI ratio in amorphous wires was explained by the particular domain structure assumed for the nearly-zero magnetostrictive amorphous wires consisting of an outer domain sheath with high circumferential permeability and outstanding magnetic softness [51,52].

Therefore amorphous wires with high circumferential permeability proved to be very favourable for highest GMI effect [2, 8, 51,52]. Such domain configuration is typical for the nearly-zero magnetostrictive amorphous wires mainly produced by Unitika LTD [2, 51,52]. On the other hand, these amorphous wires with vanishing magnetostriction constant present best magnetic softness [2,51, 52, 160-169]. It has been clearly demonstrated by Y. Yoshinaga et al [169] by chemical etching of 30 μm $(Co_{0.94}Fe_{0.06})_{72.5}Si_{12.5}B_{15}$ wire in 10% HNO_3, that the removing of the external layer with high transverse permeability results in simultaneous degradation of its magnetic softness, decrease of MI effect ($\Delta E_w/E_w$) and increase of the squareness ratio and coercivity. In addition, it was recently demonstrated that the applied tensile stresses can significantly modify the magnetoimpedance response of conventional amorphous wires and glass coated microwires [164-166, 170-172].

As has been mentioned, initially the GMI effect was interpreted in terms of the classical skin effect in a magnetic conductor assuming scalar character for the magnetic permeability, as a consequence of the change in the penetration depth of the *ac* current caused by the *dc* applied magnetic field. The electrical impedance, Z, of a magnetic conductor in this case is given by [51, 52, 173]:

$$Z = R_{dc}\, kr\, J_o(kr) / 2\, J_1(kr) \tag{50}$$

with $k = (1+j)/\delta$, where J_o and J_1 are the Bessel functions, r –wire's radius and δ the penetration depth given by:

$$\delta = (\pi \sigma \mu_{\phi} f)^{-1/2} \tag{51}$$

where σ is the electrical conductivity, f the frequency of the current along the sample, and μ_{ϕ} the circular magnetic permeability assumed to be scalar. The *dc* applied magnetic field introduces significant changes in the circular permeability, μ_{ϕ}. Therefore, the penetration depth also changes through and finally results in a change of Z [51, 52, 173]. Recently this "scalar" model was significantly modified taking into account the tensor origin of the magnetic permeability and magneto impedance [170, 173-175].

Usually GMI effect is characterized by the magnetoimpedance ratio, $\Delta Z/Z$, has been defined as:

$$\Delta Z/Z = [Z(H) - Z(H_{max})] / Z(H_{max}) \tag{52}$$

where H_{max} is the maximum *DC* longitudinal magnetic field of the order of few kA/m, usually supplied by a long solenoid and/or Helholtz coils.

It is worth mentioing that the general characteristic of magnetic materials with GMI effect is their outstanding soft magnetic behaviour when magnetized in the circular (transverse) direction. Consequently, appearance of the GMI effect and magnetic field dependence of impedance are intrinsically related with the magnetostriction constant, domain structure and abovementioned outstanding magnetic softness of magnetic materials. Therefore first we´ll briefly review studies of the magnetostriction constant and its relation with GMI effect.

4.3.1. Magnetostriction

As most of the magnetoimpedive samples are amorphous ferromagnets, the main parameter conditioning the magnetic softness is related with magnetoelastic energy determined by the the saturation magnetostriction constant, internal stresses and chemical composition [161- 163]. Their domain structure is then controlled by the magnetoelastic anisotropy induced by fabrication. The typical GMI material is a $Co_{94}Fe_6$–base amorphous alloy with vanishing but negative magnetostriction, λ_s, (-10^{-7}), which displays a large value of the effective permeability. In the case of wires, the domain structure is characterised by a multidomain inner nucleus with roughly axial easy direction, and an outer shell with transverse domains [2, 51, 52, 164].

However, the geometry of the system has a strong influence on the overall GMI behaviour. In particular, the cylindrical shape of wires and tubes minimizes the formation of magnetic poles at surfaces and closure domain structures, naturally favouring the softness of the circumferential magnetization process [2, 51, 52].

Nevertheless, only a few papers have been reported an experimental results on the saturation magnetostriction constant measurements [7]. On the other hand, indirect methods for evaluating the magnetostriction constant as the small angle magnetization rotation (SAMR) method [8-10] or the stress dependence of initial permeability [11] have been successfully applied mainly to low magnetostriction ($\lambda_s \approx 10^{-7}$ to 10^{-8}) metallic glasses.

Finally, the stress dependence of the GMI effect has been also employed to evaluate the magnetostriction constant of wire shaped samples [11].

For magnetic materials with vanishing magnetistriction constant and preferably magnetization rotation dominated remagnetization proceess the small angle magnetization rotation (SAMR) method [176-185] has been successfully applied to estimate magnetostriction constant. SAMR method is an indirect method for evaluating the magnetostriction constant as well as the stress dependence of initial permeability [181] have been successfully applied mainly to low magnetostriction ($\lambda_s \approx 10^{-7}$ to 10^{-8}) metallic glasses. Few attempts to evaluate the magnetostriction constant in Co-rich amorphous microwires exhibiting inclined hysteresis loops, i.e. with the magnetization rotation mechanism of the magnetization reversal have been performed [180, 182-185]. It should be mentioned, that attempts to apply SAMR method to estimate λ_s in microwires with rectangular hysteresis loops have been realized too. In this case the magnetostriction constant has been evaluated from stress dependence of magnetic susceptibility given by eq. (17)

In this case the magnetization rotation only of outer transversally magnetized shell contributes to the initial magnetic susceptibility [181, 182]. The condition for this case is the existence of the outer shell, where the magnetization is perpendicular to the external field. As regarding the hysteresis loop of such microwires, it exhibits two regions: for small applied magnetic fields the magnetization process is dominated by domain wall displacements within the inner core, while for larger applied magnetic fields the magnetization process takes place by magnetization rotation in the outer shell.

For the case of rotation, the magnetic energy of the microwire can be written as:

$$E = -\mu_0 M_s H_z \cos\alpha + K_{me} \sin^2(\alpha) \tag{53}$$

where α is the angle between the microwire axis and the local magnetic moment, K_{me} the magnetoelastic anisotropy introduced by the applied tensile stress, σ, and by internal stresses, σ_{oc}. Then, $K_{me} = (3/2)\lambda_s(\sigma-\sigma_{oc})$. The equilibrium condition is $\partial E/\partial\alpha = 0$. For $\sigma<\sigma_{oc}$, we get

$$\mu_0 M_s H_z + 3\,\lambda_s(\sigma-\sigma_{oc})\cos\alpha = 0 \tag{54}$$

where, considering that $\cos\alpha = M_z/M_s$, the susceptibility $\chi = \partial M_z/\partial H_z$ and that the magnetization rotation occurs only within the outer shell (with the volume, $V \approx 1-m_r$, where $m_r = \mu_0 M_r/\mu_0 M_s$) it becomes

$$\lambda_s = -1/3\,\mu_0 M_s^2\, d\chi^{-1}/V d\sigma_{appl} \tag{55}$$

which is the same expression obtained in ref. [176] for amorphous ribbons with negative magnetostriction.

Additionally, stress dependence of the GMI effect can be used for magnetostiction constant estimation, since displacement of the magnetic field corresponding to a maximum of $\Delta Z/Z$ under applied stress is usually observed. In according to well established point of view the magnetic field, H_m, at which $\Delta Z/Z(H)$ maximum takes place is attributed to magnetic anisotropy field. Accordingly in agreement with eq. 15, the magnetostriction constant can be estimated from the stress dependence of H_m [180]. Comparison of λ_s values estimated from SAMR method and from stress dependence of GMI effect gives reasonable agreement [180].

The magnetostiction constant has been evaluated for as-prepared and heat treated microwires (see tables 2 and 3).

The evolution of λ_s with the annealing time (the current, *i*, as parameter) for selected glass-coated amorphous microwires are presented in Figs. 37 (a-d). As can be observed, all samples exhibit negative λ_s ranging between $-0.9 \bullet 10^{-6}$ and $-0.3 \bullet 10^{-6}$. Under the effect of Joule heating, increasing both, t_{ann} or *i*, λ_s tends to increase approaching to zero (see Figs. 37 a-d). General tendency of increasing of λ_s with t_{ann} and *i* should be attributed to the relaxation of the internal stresses during the annealing and consequently (see eq. (18)) with the stress dependence of the λ_S. Such as has been mentioned, strong internal stresses have been predicted in glass coated microwires due to a significant difference between the thermal expansion coefficients of the metallic nucleus and the glass coating [2-4]. Therefore one can expect an increase of the magnetostriction constant after annealing.

Similarly to the magnetostrictive behaviour widely reported [102, 179] for amorphous ribbons and conventional wires of similar compositions, rather low magnetostriction values exhibited by annealed microwire samples could be explained assuming the existence of two kind of short ordering, namely: an ordered highly anisotropic phase centered at Fe atoms with uniaxial shape and magnetic anisotropy and with positive magnetostriction, mainly due to anisotropic exchange mechanism and a disordered phase at Co-rich atoms with negative magnetostriction. If we also assume that the positive magnetostriction phase (Fe atoms) undergoes a first order transformation from the uniaxial structural units (low temperature phase) towards a more symmetric and less anisotropic structural units (high temperature phase), this allows us to explain the general behaviour of macroscopic average magnetostriction, λ_s^{av}, of such Co-rich alloys. In fact, for pure Co the magnetostriction at room temperature is $- 4 \bullet 10^{-6}$ and evolves with the temperature as expected from the single ion model. Therefore, the changes of λ_s^{av} with the current annealing treatment repeated in this work on glass-coated amorphous microwires are assumed to be a consequence of the atomic rearrangements of topological character associated with both structural and stress relaxation processes. These processes would produce relative changes of the two mentioned volume fractions, so affecting the λ_s^{av}. In this way some peculiarities observed in Figs. 37 (b-d) can be roughly understood. It is worth mentioning that the high value of the internal stresses of these glass-coated microwires makes that the thermal treatments can be considered as a stress annealing and, consequently, could develop an uniaxial magnetic anisotropy [56], which should play an important role on the magnetostrictive behavior. Particularly, the induction of a strong magnetic anisotropy at low annealing temperature has been observed in CoMnSiB microwires under the effect of applied axial magnetic field. Such peculiar behaviour was interpreted as reinforcing effect of mentioned high internal stresses during field annealing of Co-rich microwires [56].

Such as reported above, the magnetostriction constant can be also estimated from the stress dependence of GMI ratio.

Table 2. Compositions and magnetostiction of Co-rich microwires

Composition	λ_s	Ratio t/d
$(Co_{1-x}Mn_x)_{75}Si_{10}B_{15}$ (x=0.09)	-3.5 10^{-7}	0.51
$(Co_{1-x}Mn_x)_{75}Si_{10}B_{15}$ (x=0.10)	4.5 10^{-7}	0.431
$(Co_{1-x}Mn_x)_{75}Si_{10}B_{15}$ (x=0.105)	6 10^{-7}	0.35
$Co_{56.5}Fe_{6.5}Ni_{10}B_{16}Si_{11}$	-0.3 10^{-7}	0.425

Table 3. Values of $\lambda_{S,0}$ and the coefficient estimated for the microwire $(Co_{1-x}Mn_x)_{75}Si_{10}B_{15}$ (x = 0.09) by the different experimental methods

Microwire $(Co_{1-x}Mn_x)_{75}Si_{10}B_{15}$ (x=0.09)	Measurement method	$\lambda_{S,0}$	A $(MPa)^{-1}$
As-cast	$(\chi_i)^{-1}(\sigma)$	$-2.1 \cdot 10^{-7}$	$11 \cdot 10^{-10}$
Field annealing	GMI(σ). (f = 2 MHz)	$-0.4 \cdot 10^{-7}$	$23 \cdot 10^{-10}$
Field annealing	GMI(σ). (f = 10 MHz)	$-1.1 \cdot 10^{-7}$	$14 \cdot 10^{-10}$

Stress dependence of GMI has been studied in $Co_{68.5}Mn_{6.5}Si_{10}B_{15}$ microwire annealed at 100 ºC for 1 hour. This GMI effect is extremely sensitive to the application of external tensile stresses, σ: magnetic field, H_m corresponding to the maximum of *($\Delta Z/Z$)* shows a roughly linear increase with σ (Figure 38). The origin of this dependence should be related with the change of the domain structure inside the outer shell with circular orientation.

It is remarkable to mention that according to ref. [2, 51, 52], the value of the *dc* axial field that corresponds to the maximum GMI ratio, H_m, should be attributed to the static circular anisotropy field, H_k. This argument allows us to estimate the magnetostriction constant using the dependence $H_m(\sigma)$ presented in the Figure 38 and the well known expression for the stress dependence of anisotropy field, given above by eq. (15)

The $H_m(\sigma)$ dependence (see the Figure 38b) exhibit a slope of around 0.7 A/(m×MPa), that allows an estimation of the unstressed value of the saturation magnetostriction constant, $\lambda_{s,0}$. Estimated value of $\lambda_{s,0} \approx -2\times10^{-7}$ have been obtained, which are rather reasonable in comparison with the reported values measured from the stress dependence of initial magnetic susceptibility ($\lambda_{s,0} \approx -3\times10^{-7}$ for such composition) [2,174-178] (see table 3).

4.3.2. GMI Effect in Different Families of Wires

The GMI effect is related intrinsically with the hysteresis loops of the samples. Typical hysteresis loops of 3 families of studied wires are presented in Figure 39.

As can be observed, Fe- and Co-rich wires have rectangular hysteresis loops, Co-Fe-rich wires exhibit enhanced magnetic softness (Figure 39a), cold-drawn wires exhibit enhanced coercivity (Figure 39c). The chemical composition of glass-coated microwires drastically affects their magnetization curves: the hysteresis loop changes from rectangular for Fe-rich compositions to almost unhysteretic in Co-rich compositions (Figure 39b and Figure 7).

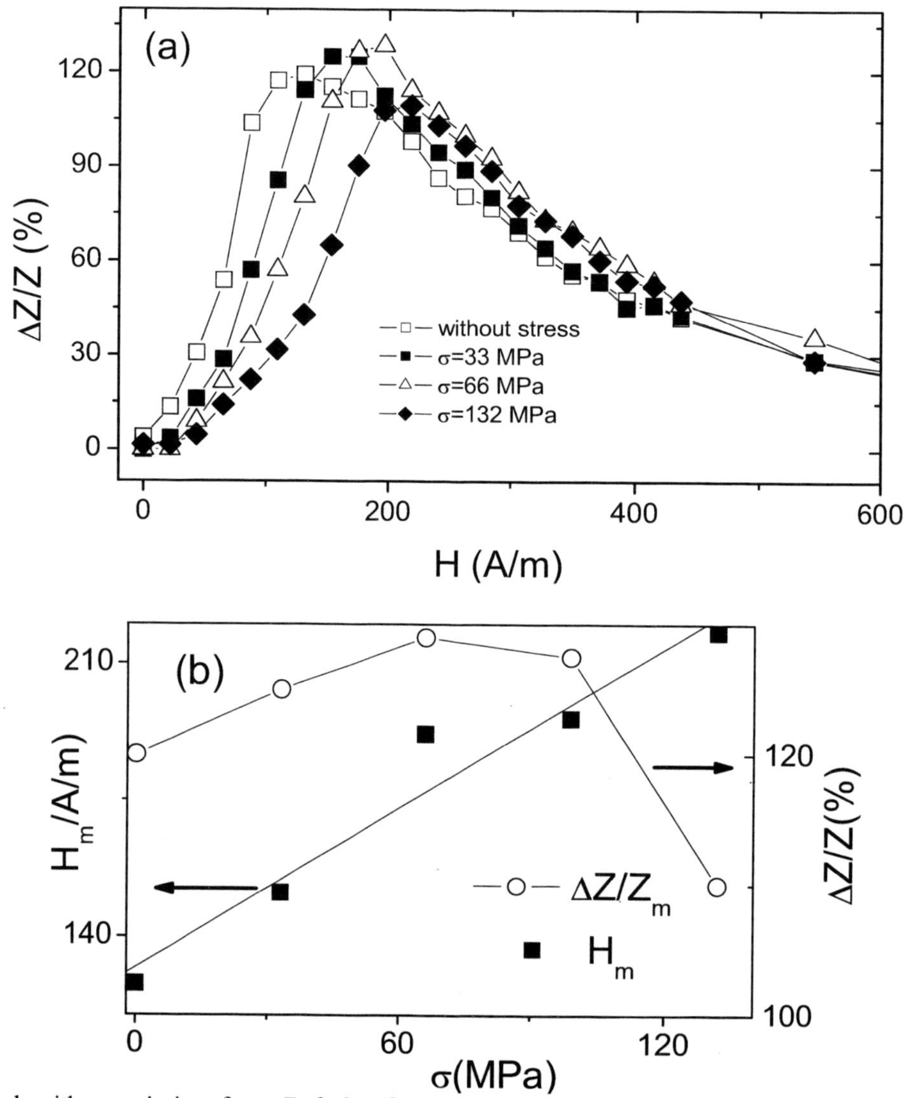

Reprinted with permission from Ref. [136], A. Zhukov, "Glass-coated magnetic microwires for technical applications" J. Magn and Magn, Mater., 242-245, p216 (2002). Copyright Elsevier (2002) (figure 8a).

Figure 38. *ΔZ/Z(H)* dependencies of $Co_{68.5}Mn_{6.5}Si_{10}B_{15}$ amorphous microwire measured at different σ (a) and H_m (σ) dependence (b).

The magnetic field, *H*, dependence of the magneto-impedance ratio, *ΔZ/Z*, measured at different frequencies, *f*, for as-cast $Fe_{77.5}B_{15}Si_{7.5}$ and $(Co_{0.94}Fe_{0.06})_{72.5}Si_{12.5}B_{15}$ amorphous conventional wires are presented in the Figure 40. In $Fe_{77.5}B_{15}Si_{7.5}$ amorphous wires between 10 MHz and 100 MHz the GMI effect presents monotonic decay of the impedance with magnetic field, while increasing the frequency above 100 MHz such dependence changes the character showing a maximum at certain *dc* axial magnetic field, which depends on the frequency (see Figure 40).

On the other hand Fe-rich cold drawn wires show a maximum on the field dependence of the impedance, Z, even 10 MHz and considerable hysteresis (see Figure 41).

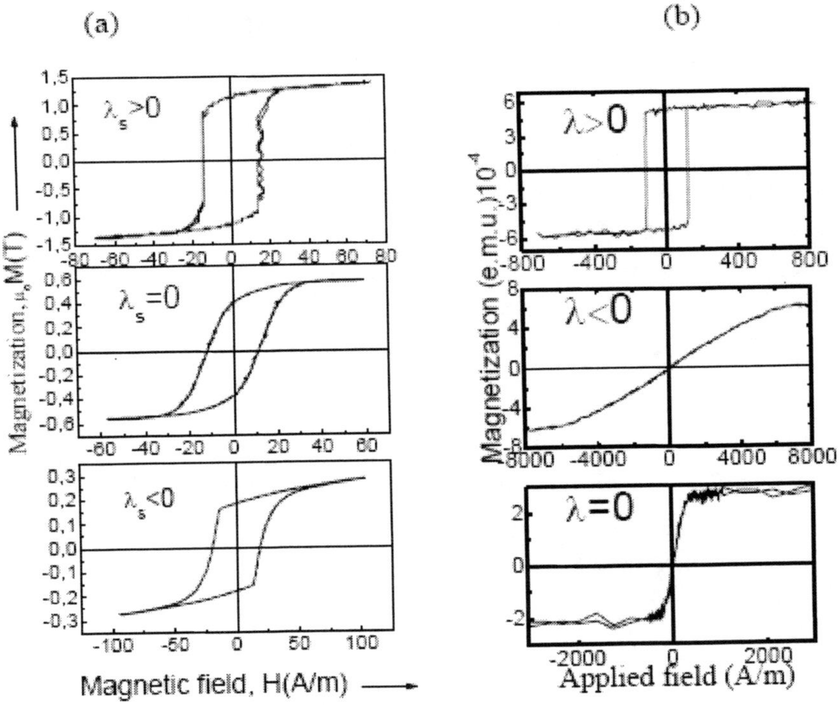

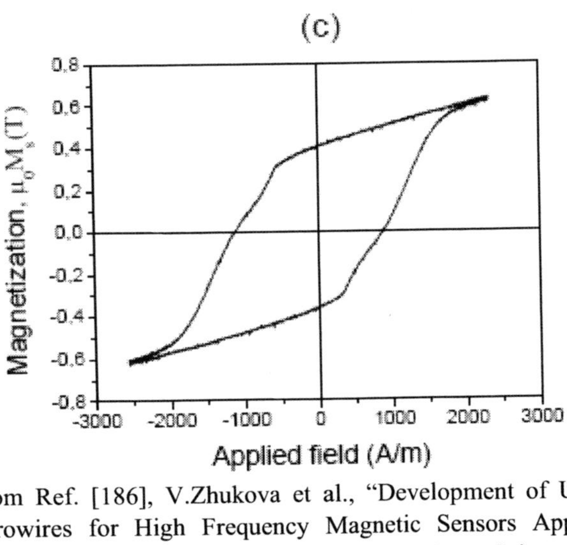

Reprinted with permission from Ref. [186], V.Zhukova et al., "Development of Ultra-Thin Glass-Coated Amorphous Microwires for High Frequency Magnetic Sensors Applications" Open Materials Science Reviews, 2007, 1, 1-12. Copyright 2007 Bentham Science Publishers Ltd. (figure 1).

Figure 39. Hysteresis loops of (a) conventional amorphous wires, (b) thin glass coated microwires and (c) Fe-rich cold drawn wires.

Finally magnetic field dependence of the impedance has been also measured in 3 different thin glass-coated microwires at the same frequency range (10-500 MHz) (see Figure 42). At 10 MHz the GMI of Fe-rich microwires is smaller than in cold-drawn wires, but increasing the frequency GMI effect significantly increases. The shape of the Z(H) is quite different from cold drawn wires, showing roughly only the decay with DC applied magnetic field. Small maximum can be appreciated at about 500 MHz. Co-rich microwires exhibit much higher GMI effect at all frequencies and the shape of the Z(H) dependence is typical for the materials with circular magnetic anisotropy, i.e. with the a maximum at certain *dc* axial magnetic field.

The main difference in Z(H) dependences for 3 different families of amorphous wires can be summarized as following: i) GMI ratio decreases with frequency between 10 and 500 MHz in conventional wires (for Fe-rich wires this decreasing follows after first increasing of $\Delta Z/Z_{max}$), while increases in thinner wires (both cold drawn and glass coated); ii) Cold drawn wires possess considerable hysteresis on Z(H) dependences; iii) Even Fe-rich cold amorphous wires exhibit considerable GMI effect at elevated frequencies.

These results are summarized in Figs. 43 and 44 where $\Delta Z/Z_{max}$ (f) and H_m(f) dependences are shown.

A remarkable difference in magnetic field dependence of the GMI effect should be attributed to the different magnetoelastic anisotropy of these three families of the wires. Thus, as-cast conventional wires possess smallest internal stresses of the thermal origin. In the case of cold drawn wires the additional strong stresses are arising from the cold drawn processing. These cold drawn wires possess enhanced magnetic hardness, as-comparing with the other families of studied wires. Therefore the magnetic anisotropy field, H_m, associated with the field, at which the maximum on the Z(H) dependence takes place [186], is the highest. Alternatively, glass-coated wires are the composite materials. Consequently the glass-coating technology gives rise to the internal stresses due to the difference in the thermal expansion coefficients of the glass coating and metallic nucleus. These differences in the fabrication technique result in the different magnetic anisotropy in the surface and in different frequency dependence of the GMI effect. On the other hand quite unusual frequency dependence of the H_m is observed in cold drawn Fe-rich wires (Figure 44c). In fact this is the unique wire showing non-monotonic dependence of H_m (f). Such non-monotonic dependence can be attributed to the cold drawn induced magnetoelastic anisotropy in the surface layer of the wire.

4.3.3. Tailoring of Hysteretic Properties and GMI Effect in Thin Glass-coated Microwires Using Heat Treatment. Induce Magnetic Anisotropy. Stress Sensitive Microwires

As has been being observed, hysteresis loops are quite sensible to the chemical composition and sample geometry (see Figs, 7, 8, 39). While Co-rich compositions present roughly inclined hysteresis loop in as-prepared state, Fe-rich microwire exhibit perfectly rectangular hysteresis loop.

Such strong dependence of the hysteresis loops on these parameters can be related with the magnetoelastic energy. In fact the magnetoelastic energy is given by above by eq.14.

The magnetostrction constant determines mostly by the chemical composition achieving nearly-zero value in amorphous Fe-Co based alloys al about %Co/%Fe ≈70/5 [2]. On the other hand, large internal stresses give rise to a drastic change of the magnetoelastic energy, K_{mei}, even for small changes of the glass-coating thickness at fixed metallic core diameter.

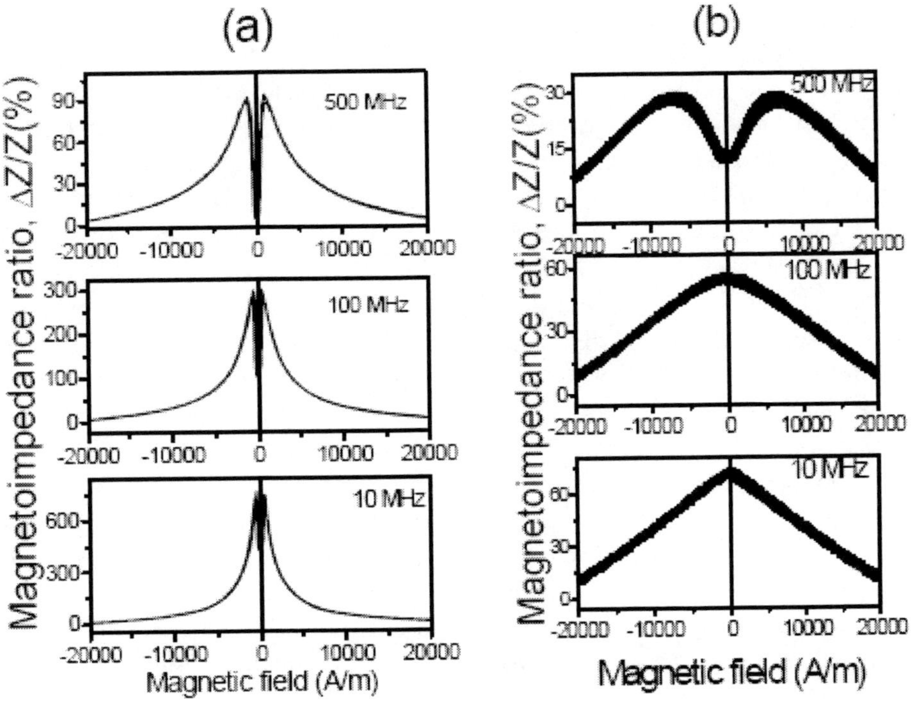

Reprinted with permission from Ref. [186], V.Zhukova et al., "Development of Ultra-Thin Glass-Coated Amorphous Microwires for High Frequency Magnetic Sensors Applications" Open Materials Science Reviews, 2007, 1, 1-12. Copyright 2007 Bentham Science Publishers Ltd. (figure 2).

Figure 40. DC magnetic field dependence of the impedance of the conventional $(Co_{0.94}Fe_{0.06})_{72.5}Si_{12.5}B_{15}$ (a) and $Fe_{77.5}B_{15}Si_{7.5}$ (b) amorphous wire with diameter of 125 μm.

These considerations allows us to predict that any method to change the internal stresses (by using thermal treatment, chemical etching etc) can change drastically magnetic anisotropy and consequently the hysteresis loops and the GMI behaviour.

Consequently the hysteresis loops of $Co_{69-x}Mn_{6+x}Si_{10}B_{15}$ (x=0.5) has been changed after conventional annealing, CA, or under magnetic field current annealing, MFA treatments (see Figure 45).

In the case of MFA, the external axial magnetic field applied during the MFA treatment induces axial magnetic anisotropy, such as can be appreciated from Figure 45b. In fact during current annealing a circular magnetic field has been also, applied. Therefore, although current annealing is the easiest way to heat treat the samples, conventional furnace annealing gives more clear ideas on the induction of magnetic anisotropy by magnetic field annealing.

Dependencies of coercivity on the annealing time, t_{ann}, for conventional current annealing (CA) and magnetic field current annealing (MFA) with current density 227 A/mm^2 for three microwires are shown in Figure 46. In the case of the $Fe_{70}B_{15}Si_{10}C_5$ microwire, both treatments (CA, FA) at 30 mA ($j=227\ A/mm^2$) caused a decrease of H_c with t_{ann}. In the case of MFA, such decrease of coercivity is more significant. In contrast, in the case of $Co_{56,5}Fe_{6,5}Ni_{10}B_{16}Si_{11}$ samples, both treatments result in opposite effects [17]. CA treatment gives rise to a roughly monotonic decrease of the coercivity, while MFA results in an increase of H_c, with a broad maximum at t_{ann}≈10 min. Similarly to $Fe_{70}B_{15}Si_{10}C_5$ samples, a monotonic decrease of H_c has been observed for $Co_{68}Mn_7Si_{10}B_{15}$ sample after both CA and MFA treatments. Besides, as in the case of the $Co_{56,5}Fe_{6,5}Ni_{10}B_{16}Si_{11}$ sample, the MFA current annealing treatment results in higher H_c, as comparing with the CA treatment.

Similarly to current annealing, conventional furnace annealing resulted in drastic changes of hysteretic properties under application of magnetic field during annealing.

Series of different treatments up to 473 K, such as conventional annealing (CA), magnetic field annealing (MFA) have been performed in the samples $(Co_{1-x}Mn_x)_{75}Si_{10}B_{15}$ with x=0.08 [56,87]. Figures 47 (a) show as an example the hysteresis loops obtained for the sample x=0.08 after heat treatments at 373, 423 and 473 K. The conventional low-temperature thermal treatments (CA) (in the range of 373 – 473 K, 1 hour) produce small changes in the coercive field, which can be ascribed to the relaxation of internal stresses within the amorphous state. Nevertheless, when the thermal treatment was carried out under the presence of an axial magnetic field (field annealing, MFA), significant changes of the hysteresis loop were detected (see Figure 47b). This change produces an increase of the initial permeability, μ_i, coercivity, H_c, remanent magnetization, $\mu_0 M_r$, and a decrease of the anisotropy field, H_K (see Figures 48a, 48b and 48c). The changes of the shape of the hysteresis loop with FA should be connected to the induction of magnetic anisotropy with easy axis along the longitudinal direction of the microwire. This anisotropy is developed by the combined effect of the high internal stress and the magnetic field applied during the treatment [56,87].

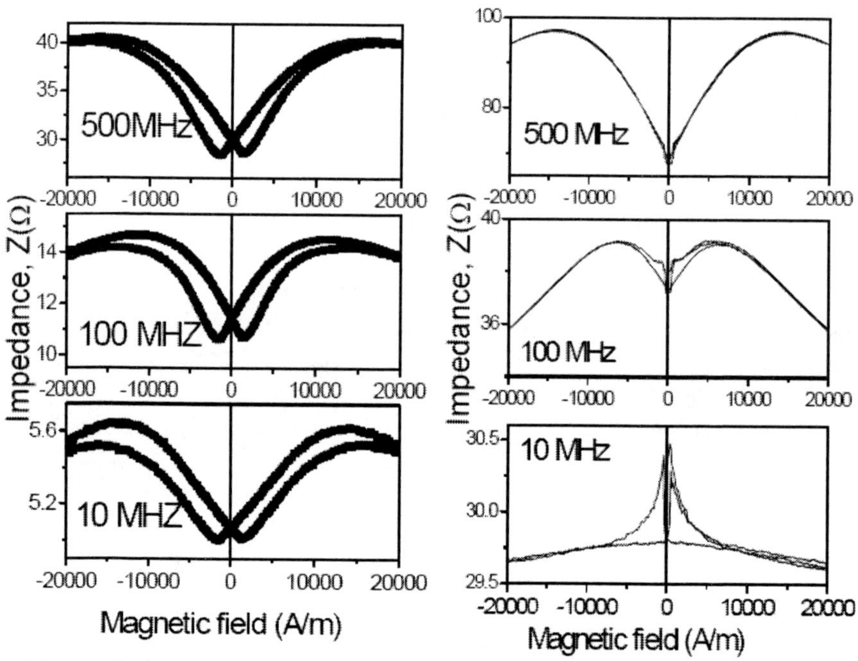

Reprinted with permission from Ref. [186], V.Zhukova et al., "Development of Ultra-Thin Glass-Coated Amorphous Microwires for High Frequency Magnetic Sensors Applications" Open Materials Science Reviews, 2007, 1, 1-12. Copyright 2007 Bentham Science Publishers Ltd. (figure 3).

Figure 41. DC magnetic field dependence of the impedance of the cold-drawn amorphous wire $Fe_{77.5}B_{15}Si_{7.5}$ with diameter of 50 μm (a) and 30 μm (b).

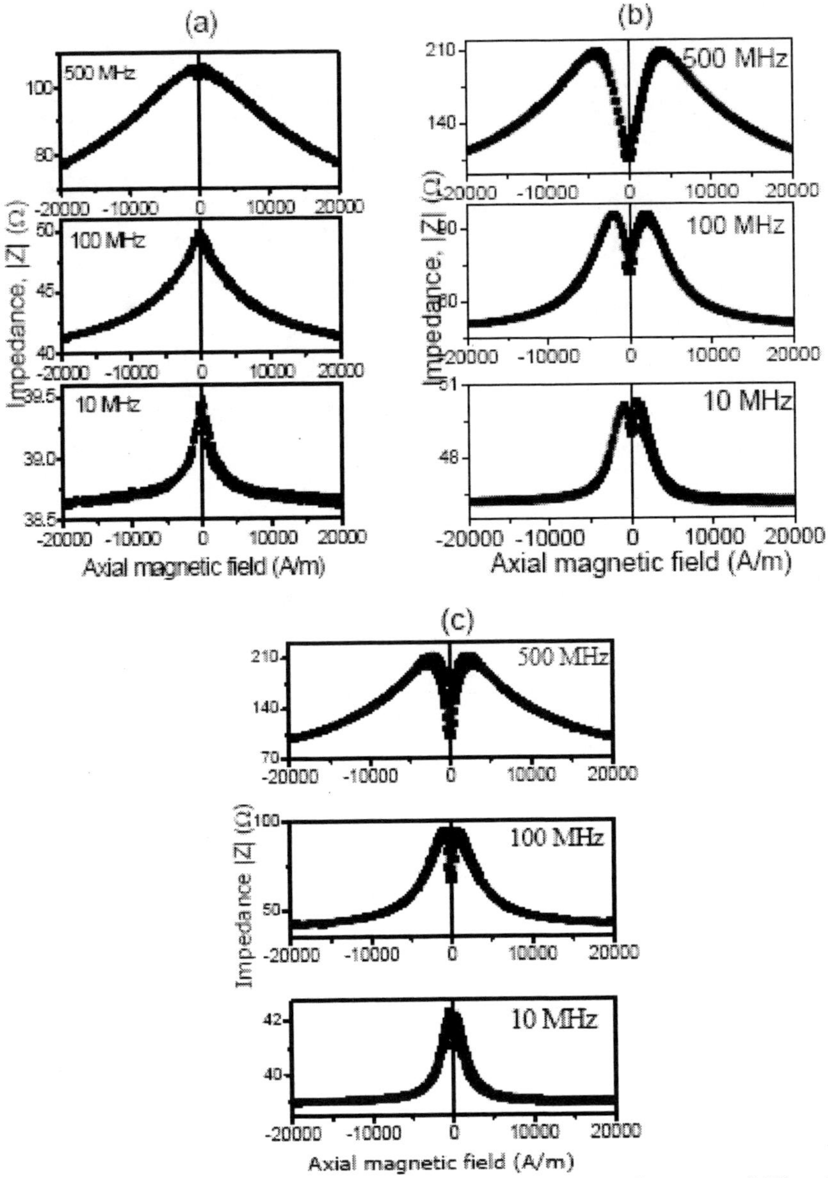

Reprinted with permission from Ref. [186], V.Zhukova et al., "Development of Ultra-Thin Glass-Coated Amorphous Microwires for High Frequency Magnetic Sensors Applications" Open Materials Science Reviews, 2007, 1, 1-12. Copyright 2007 Bentham Science Publishers Ltd. (figure 4).

Figure 42. Z(H) dependence of $Fe_{75.5}B_{13}Si_{11}Mo_{0.5}$ (a), $Fe_{3.7}Co_{69.8}Ni_1Si_{11}B_{13}Mo_{1.5}$ (b) and $Co_{77.5}Si_{7.5}B_{15}$ (c) glass-coated amorphous microwires.

The observed difference between the CA and MFA should be attributed to the effect of the stress+longitudinal field induced magnetic anisotropy in the last case, where the strong internal stresses contribute as the reinforcing factor. This feature implies that the application of the longitudinal magnetic field during annealing favors the formation of the internal axially magnetized core with longitudinal magnetic anisotropy. At the mean time, the conventional current annealing (CA) results only in internal stress relaxation [56, 87].

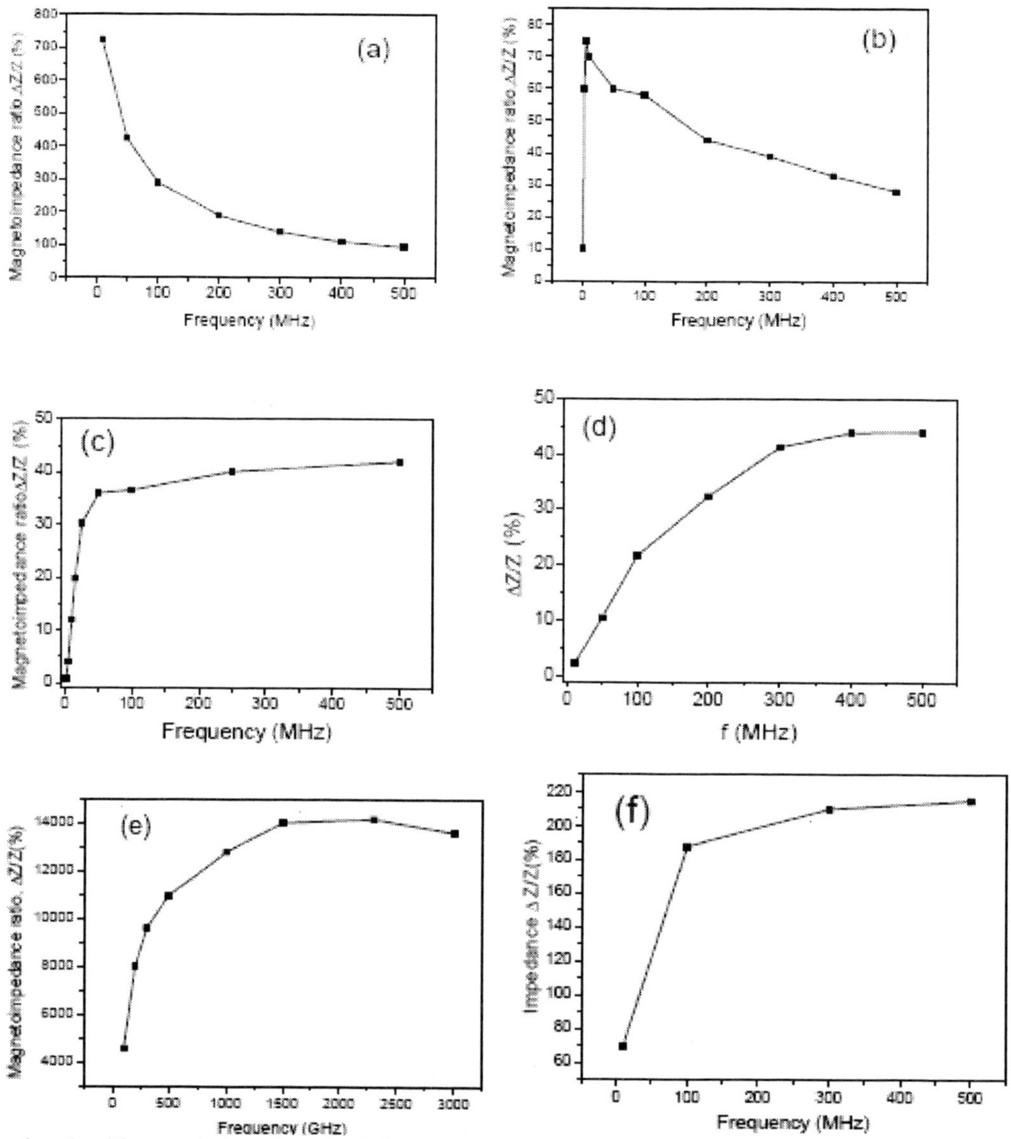

Reprinted with permission from Ref. [186], V.Zhukova et al., "Development of Ultra-Thin Glass-Coated Amorphous Microwires for High Frequency Magnetic Sensors Applications" Open Materials Science Reviews, 2007, 1, 1-12. Copyright 2007 Bentham Science Publishers Ltd. (figure 5).

Figure 43. $\Delta Z/Z_{max}(f)$ dependence for different amorphous wires: Conventional (a)- $(Co_{0.94}Fe_{0.06})_{72.5}Si_{12.5}B_{15}$; and (b) - $Fe_{77.5}B_{15}Si_{7.5}$ wires; Cold drawn (c)- $Fe_{77.5}B_{15}Si_{7.5}$ (50 μm) wires and glass-coated (d)- $Fe_{75.5}B_{13}Si_{11}Mo_{0.5}$; (e)- $Co_{77.5}Si_{7.5}B_{15}$ and (f)- $Fe_{3.7}Co_{69.8}Ni_1Si_{11}B_{13}Mo_{1.5}$ microwires.

Until know it was assumed that the Fe-rich microwires always present rectangular hysteresis loop associated with the strong axial magnetic anisotropy induced by residual stresses of mostly tensile origin [2-4]. Generally the conventional annealing of Fe-rich microwires dos not affect the character of the hysteresis loops, changing slightly the coercive force. As an example see the effect of the heat treatment performed at T_{ann}=400 °C in $Fe_{74}B_{13}Si_{11}C_2$ (see Figure 49): some decrease of H_c has been observed while the squared character of the hysteresis loop remains unchanged. On the other hand, it was recently found that the stress annealing, SA, performed at the same annealing conditions (T_{ann}=400 °C) but under applied stress, σ=458 MPa, results in drastic changes of the hysteresis loop (see Figure 49c): hysteresis loop becomes inclined with a magnetic anisotropy field about 1000 A/m.

A transverse magnetic anisotropy induced by the SA allows us to predict the existence of the magneto impedance effect in such samples. Indeed, if the samples with rectangular hysteresis loop do not present any significant GMI effect at 10 MHz, considerable GMI effect (about 13%) has been observed in stress annealed Fe-rich sample with inclined hysteresis loop (see Figure 50).

A number of heat treatments with annealing temperatures between 100 and 400°C have been performed (see Figures 51, 52) [2, 187- 189]. In fact annealing at temperatures above 400°C can deteriorate the glass coating, therefore main attention has been paid to the region T_{ann}< 400°C. Figure 51 shows the hysteresis loops of 2 samples annealed at the same temperature, T_{ann}, (275°C) with the same applied tensile stress but at different annealing time. As can be appreciated the transverse magnetic anisotropy increases increasing the SA time. Strong dependence of the transverse magnetic anisotropy reflected in the shape of axial hystresis loop on annealing temperature has been observed. As it is shown in Figs. 51 and 52, the hysteresis loop becomes more and more rectangular decreasing the annealing time (Figure 51) and decreasing the annealing temperature from 275 C to 100 °C (Figure 52).

Observed changes of the shape hysteresis loops with transverse magnetic anisotropy changing from very slight to strong enough depending on the SA annealing temperature (see Figures 52) allows us to predict the existence of the magneto-impedance effect in such samples. Indeed, if as-prepared samples do not present any GMI effect, the samples subjected to SA exhibit significant magneto-impedance (see Figure 53). As shown in Figure 53, significant GMI effect (up to 60%) is observed in annealed under stress samples. GMI ratio, $\Delta Z/Z$, increases with increasing the annealing temperature until 275 °C and finally decreases when the transverse anisotropy is too large (see Figure 53). Axial field dependences have a shape of monotonic decay from H=0 (typical for λ_s>0).

Another important result is that the application of tensile stresses to the samples subjected to the SA treatment results in the recovery of the rectangular hysteresis loop typical for the as-prepared state (see Figure 54).

Consequently, disappearance of the inclined hysteresis loop under applied stress allows as predicting the stress impedance effect, SI, in the sample subjected to SA treatment. Indeed considerable SI effect (about 60%) has been observed under application of tensile stress (see Figure 55) in microwire previously stress annealed at 270 °C.

The origin of this creep annealing induced anisotropy has been attributed to both: i) redistribution of the residual stresses during the stress annealing or ii) induced magnetic anisotropy related to the atomic ordering.

It has been assumed that after the thermal treatment under stress the longitudinal stresses in stress annealed samples become smaller than radial and tangential stresses, i.e. that the stress annealing results in re-distribution of the internal stresses and/or local microstructure of the sample in order to minimize the magnetoelastic energy in the stressed state. As a result the easy magnetization axis in stress annealed sample is aligned along the transverse direction.

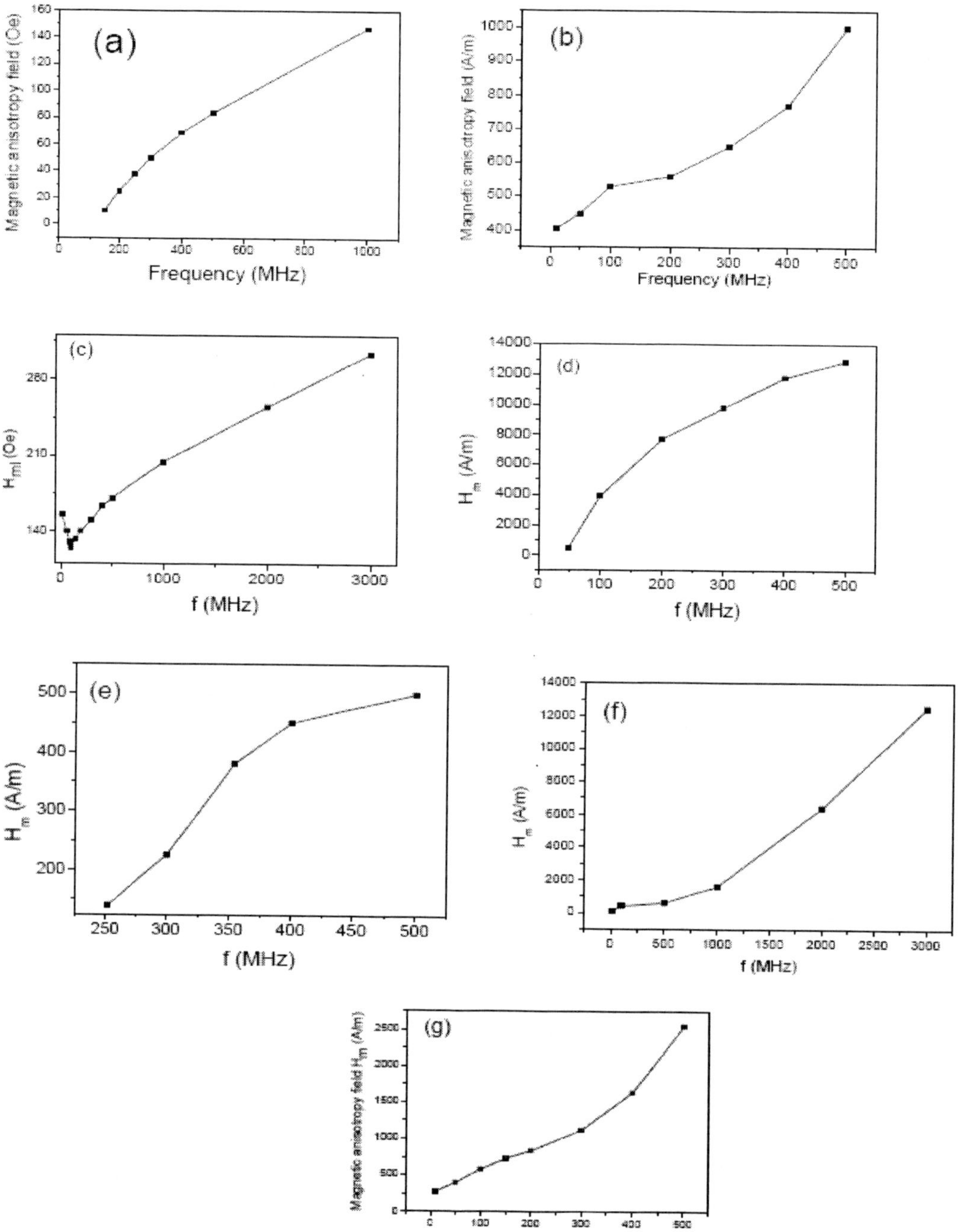

Reprinted with permission from Ref. [186], V.Zhukova et al., "Development of Ultra-Thin Glass-Coated Amorphous Microwires for High Frequency Magnetic Sensors Applications" Open Materials Science Reviews, 2007, 1, 1-12. Copyright 2007 Bentham Science Publishers Ltd. (figure 6).

Figure 44. $H_m(f)$ dependence in different amorphous wires: Conventional wires of (a) - $(Co_{0.94}Fe_{0.06})_{72.5}Si_{12.5}B_{15}$; and (b) - $Fe_{77.5}B_{15}Si_{7.5}$ compositions; Cold drawn wires (c)- $Fe_{77.5}B_{15}Si_{7.5}$ with diameter 50 μm, (d) - $Fe_{77.5}B_{15}Si_{7.5}$ with diameter 20 μm and glass-coated microwires of (e)- $Fe_{75.5}B_{13}Si_{11}Mo_{0.5}$; (f)- $Co_{77.5}Si_{7.5}B_{15}$ and (g)- $Fe_{3.7}Co_{69.8}Ni_1Si_{11}B_{13}Mo_{1.5}$ compositions.

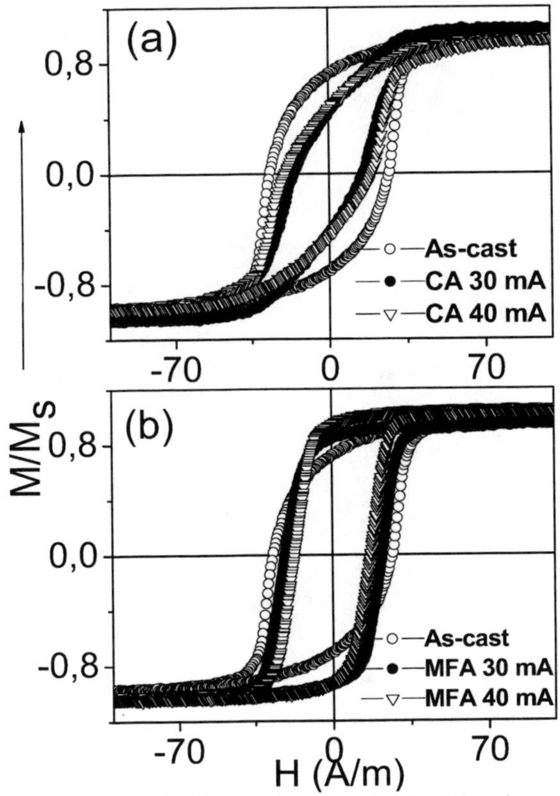

Reprinted with permission from Ref. [186], V.Zhukova et al., "Development of Ultra-Thin Glass-Coated Amorphous Microwires for High Frequency Magnetic Sensors Applications" Open Materials Science Reviews, 2007, 1, 1-12. Copyright 2007 Bentham Science Publishers Ltd. (figure 7).

Figure 45. Effect of CA (a) and MFA (b) treatments on bulk hysteresis loops of $Co_{67}Fe_{3.85}Ni_{1.45}B_{11.5}Si_{14.5}Mo_{1.7}$ microwires.

In this way, the removal of the mechanical load after stress annealing affects mostly the longitudinal stresses and results in drastic decrease of the longitudinal stress component and even in the appearance of the compressive longitudinal stresses (so-called "back stresses").

Observed change of the hysteresis loop of the sample subjected to SA treatment when measured under applied tensile stress (Figure 54) confirms such assumption. This effect should be attributed to the increasing of the longitudinal stress component under application of the tensile stress and consequently to the alignment of the easy magnetization axis along the highest stress component due to the positive magnetostriction constant of the studied sample. There is a competition between the magnetoelastic anisotropy induced by the applied stress (with a longitudinal easy axis) and the transverse anisotropy induced by the stress annealing. Upon overcoming the transverse anisotropy related to the creep induced anisotropy the longitudinal axis becomes easy magnetization axis again and magnetic bistability recovers.

In fact, under application of applied stresses to stress annealed microwire, results in considerable change of the magnetic structure and magnetic properties are observed: from inclined, the hysteresis loop become again rectangular, as in the as-prepared sample through few steps [189].

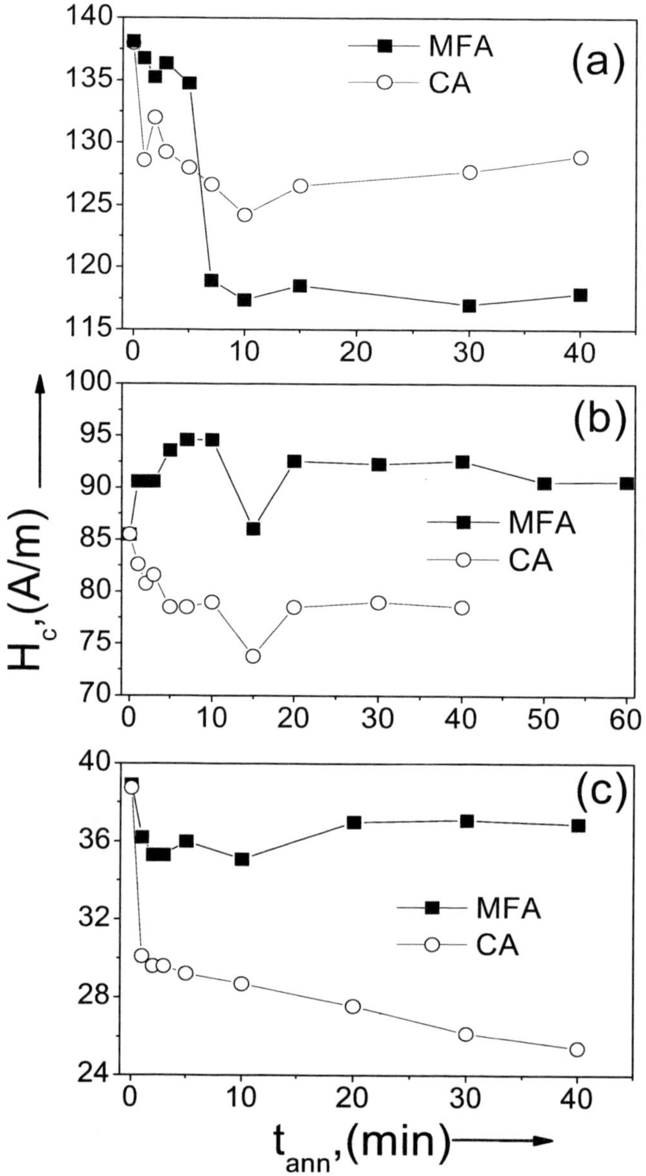

Reprinted with permission from [17] V. Zhukova et al., "Tailoring of magnetic properties of glass coated microwires by current annealing", J. Non-crystalline solids, 287, pp.31-36, Copyright 2001 Elsevier Science (figure 3).

Figure 46. Comparative evolution of H_c with annealing time for CA and FA of $Fe_{70}B_{15}Si_{10}C_5$ (a), $Co_{56,5}Fe_{6,5}Ni_{10}B_{16}Si_{11}$ and $Co_{68}Mn_7Si_{10}B_{15}$ (c) microwires annealed at $j \approx 227\ A/mm^2$ with the annealing time, t_{ann}, as a parameter.

Consequently, increasing the tensile stress, remanent magnetization and coercivity increase. Further grow of tensile stress results in appearance of hysteresis loop consisting of few Barkhausen jumps (Figure 56). Studied samples show up to 5 pulses of reversal magnetization. Under tensile stress about 500 MPA the peaks of reversal magnetization unified in one peak, with coercive force similar to as-prepared sample. All steps (totally 7 steps) are shown on Figure 56.

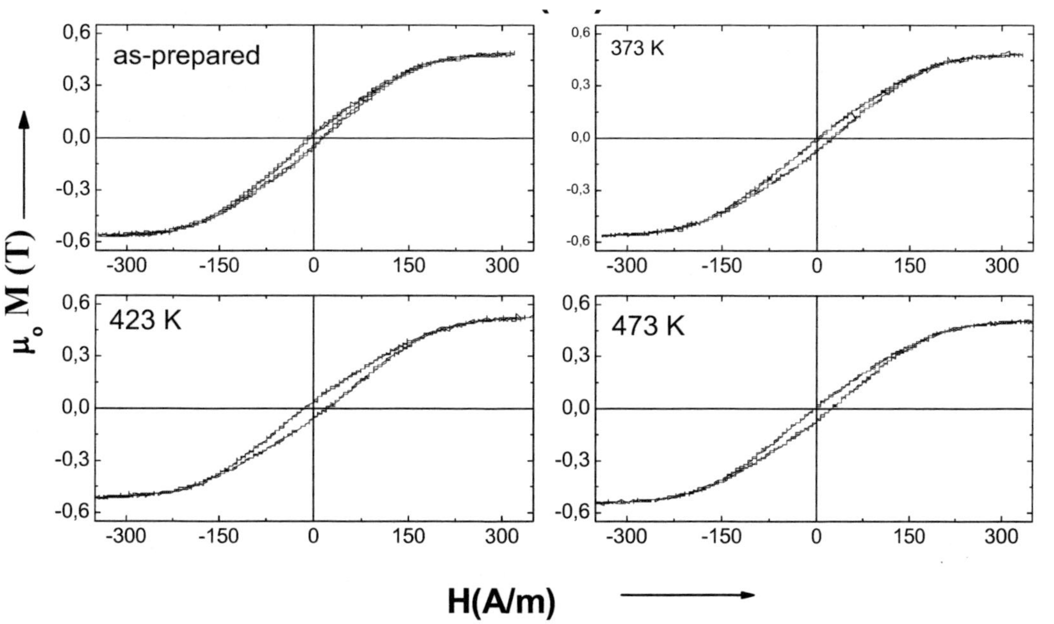

(a)

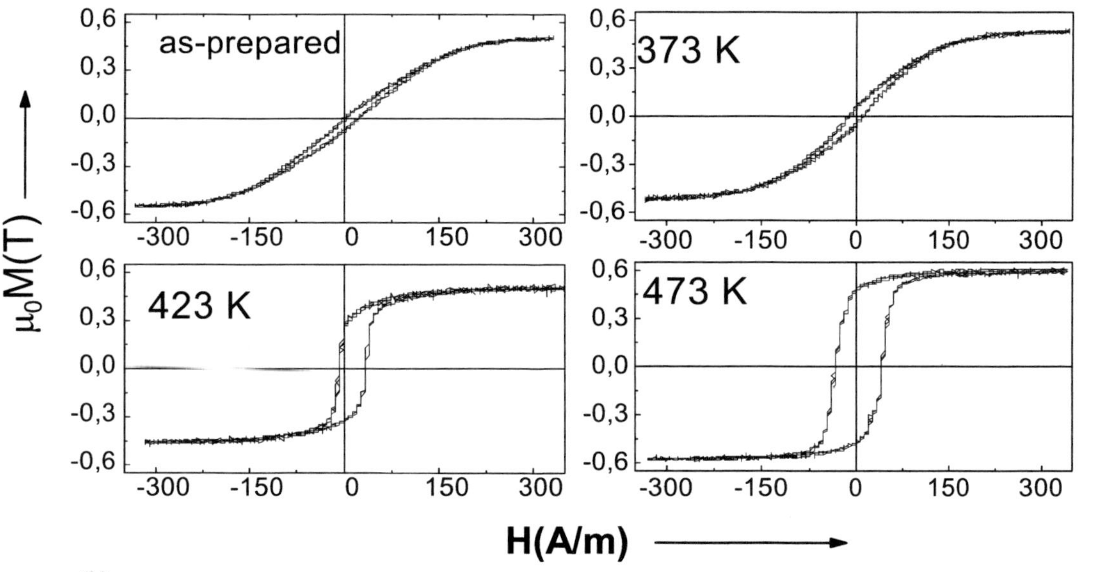

(b)

Figure 47. Effect of conventional annealing (a) and magnetic field annealing (b) on hysteresis loops of $(Co_{0.92}Mn_{0.8})_{75}Si_{10}B_{15}$ microwire at different annealing temperatures.

Changes of magnetic properties under tensile stress of microwires previously subjected TMT have been explained in the following way:

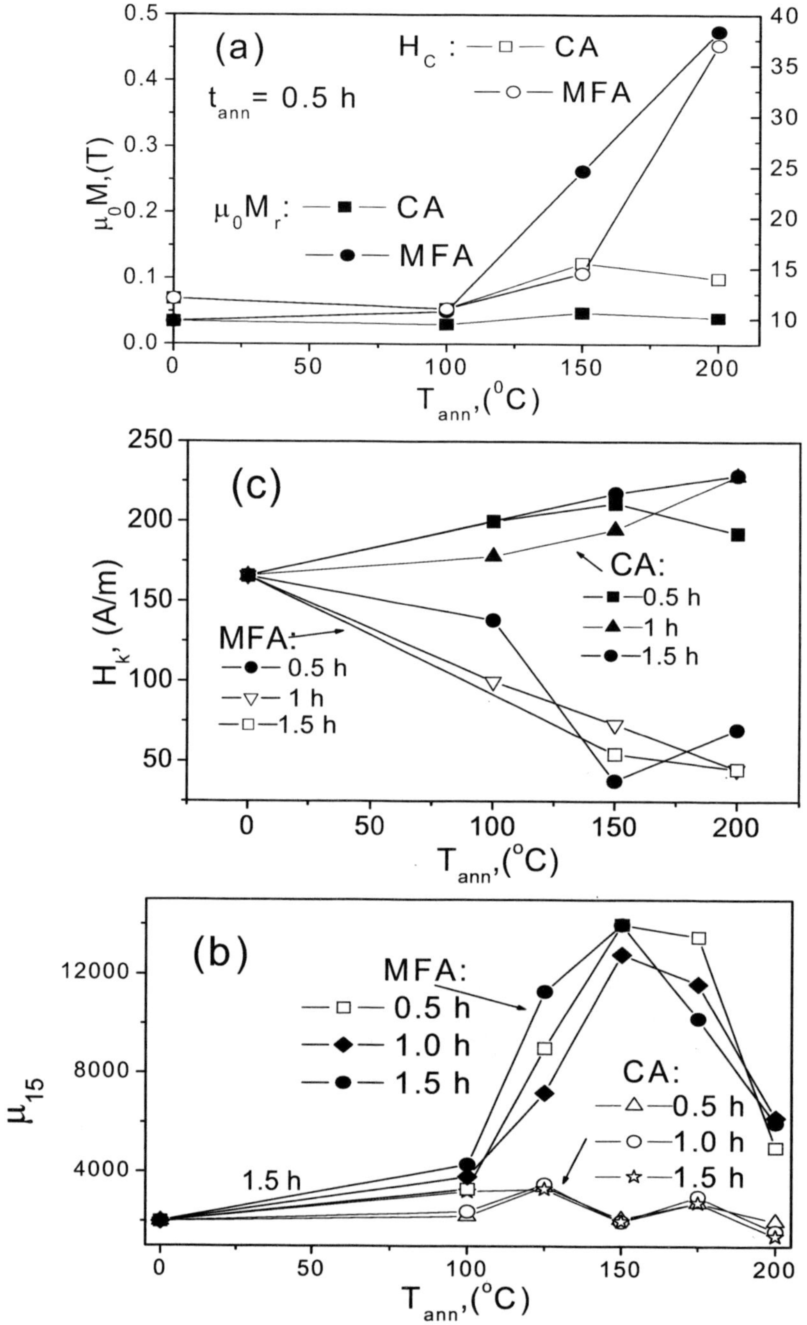

Reprinted from [36] A. Zhukov et al., "Induced Magnetic Anisotropy in Co-Mn-Si-B Amorphous Microwires", J. Appl Phys. 87 2000 1402-1408, Copyright (2000), with permission from AIP. (figure 4a,c,d).

Figure 48. Effect of magnetic field annealing (FA) on some magnetic properties of $Co_{69}Mn_6Si_{10}B_{15}$ microwires: initial magnetic permeability, μ_{15} (a), anisotropy field, H_k (b) and coercivity and remanent magnetization (c).

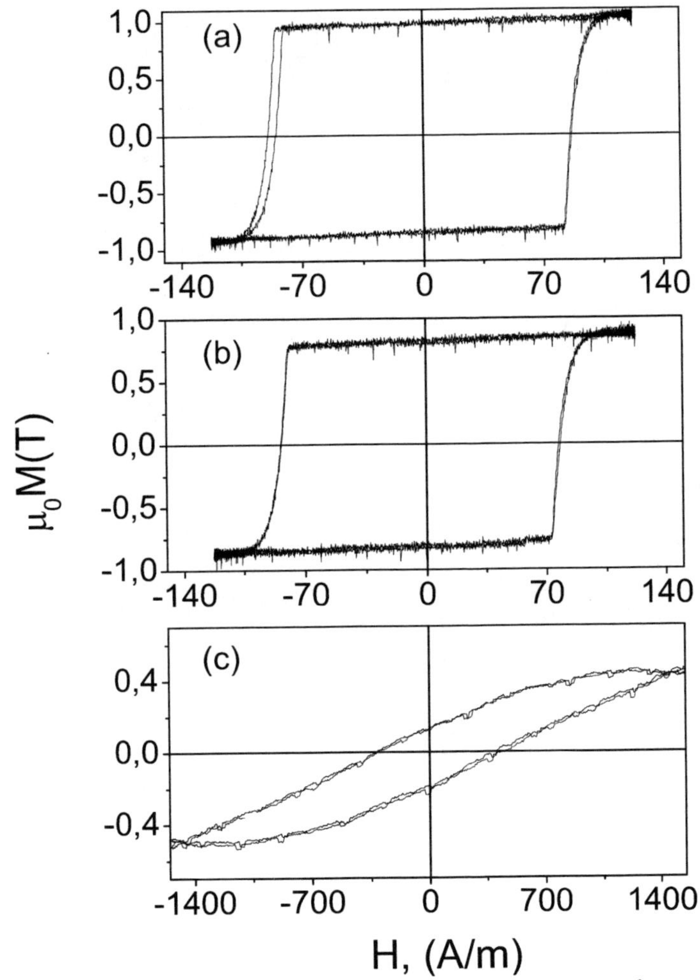

Reprinted with permission from Ref. [161], A. Zhukov et al. "Recent research on magnetic properties of glass-coated microwires" J. Magn. and Magn., Mater. 294 (2005) 182-192. Copyright Elsevier (2005) (figure 4).

Figure 49. Bulk hysteresis loops of $Fe_{74}B_{13}Si_{11}C_2$ microwire as-prepared (a), furnace annealed at T_{ann}=400°C (b) and stress annealed at T_{ann}=400°C (c).

When tensile stress is applied along microwire's axis (along axis Z) magnetoelastic component in axial direction (Z component) grows. When values of axial and radial anisotropy are similar, the resultant magnetic anisotropy field takes low values. The external layers of metallic nucleus remain magnetized circularly and have high values of $H_к$, and easy magnetization axis of the inner core has the magnetization angle 0<φ< 90° relative to the microwire axis (first stage).

When increasing the tensile stress the core of microwire tends to magnetize mainly along the wire´s axis. Coercive force can take quite low values (few A/m). Such low coercivity can be explained by the absence of structural factors that have an influence on Hc (absence of defects and heterogeneity) and compensated anisotropies along axis Z.

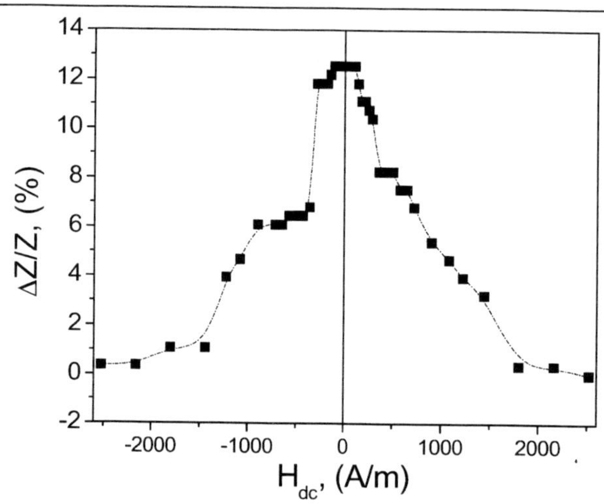

Reprinted with permission from Ref. [161], A. Zhukov "Recent research on magnetic properties of glass-coated microwires" J. Magn. and Magn., Mater. 294 (2005) 182-192. Copyright Elsevier (2005) (figure 12).

Figure 50. GMI effect induced by stress annealing of Fe-rich microwire at T_{ann}=400°C.

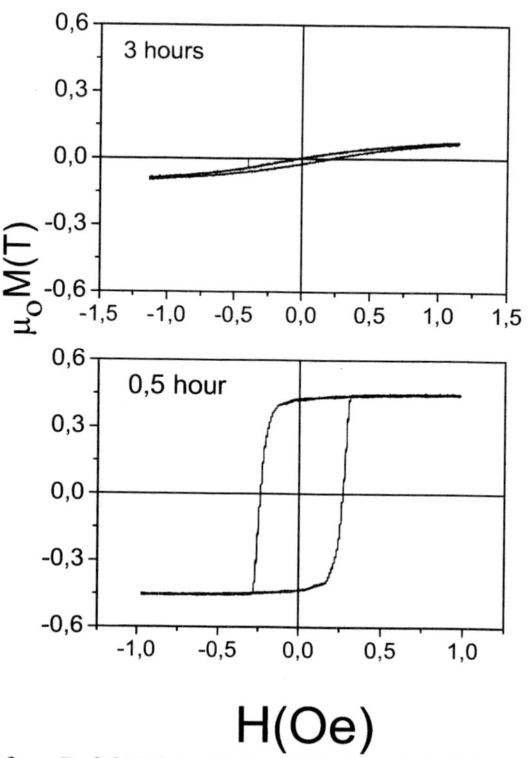

Reprinted with permission from Ref. [187] A. Zhukov "Design of the Magnetic Properties of Fe-Rich, Glass-Coated Microwires for Technical Applications" Adv. Funct. Mater. 2006, 16, 675–680. Copyright WILEY-VCH (2006) (figure 2)

Figure 51. Hysteress loops of $Fe_{74}B_{13}Si_{11}C_2$ glass-coated microwire subjected to stress annealing at 275°C for 0.5 and 3 hours.

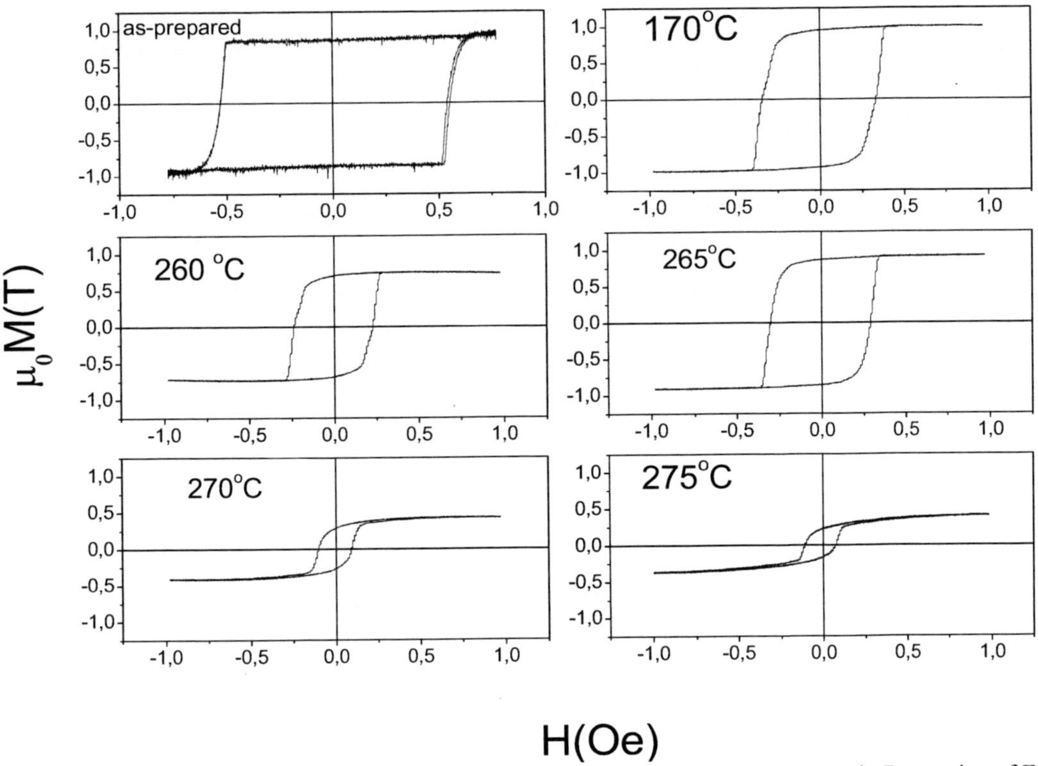

Reprinted with permission from Ref. [187] A. Zhukov et al. "Design of the Magnetic Properties of Fe-Rich, Glass-Coated Microwires for Technical Applications" Adv. Funct. Mater. 2006, 16, 675–680. Copyright WILEY-VCH (2006) (figure 3).

Figure 52. Effect of annealing temperature on hysteress loops of $Fe_{74}B_{13}Si_{11}C_2$ glass-coated microwire subjected to stress annealing for 0.5 hour.

Further increasing of the tensile stress leads to the appearance of an intermediate layer between external shell of metallic nucleus with circular magnetization and inner part of metallic nucleus magnetized along the axis. In some place again appear "conditions" of anisotropy "compensation". This intermediate layer can be re-magnetized at magnetic fields below the field at which the inner core is re-magnetizing. Further grow of tensile stress can bring to the formation of new coaxial layers. Reversal magnetization of sample take place layer-by-layer, every layer have own value of coercive force and volume. Studied samples show up to 5 Barkhausen jumps.

4.3.4. Tailoring of the GMI Effect

Until now best soft magnetic properties and highest GMI effect have been observed in $Co_{67}Fe_{3.85}Ni_{1.45}B_{11.5}Si_{14.5}Mo_{1.7}$ microwires [160, 161]. Hysteresis loops presented in Figure 7 exhibit inclined almost unhysteretic M(H) loops with extremely low coercivities (up to 4 A/m). Magnetic anisotropy field, H_k, increases with increasing the ρ–ratio defined as d/D (d- metallic nucleus diameter, D- total microwire diameter) (Figure 57). Similarly, both GMI ratio, ΔZ/Z and the field of maximum, H_m depend on ρ–ratio exhibiting $\Delta Z/Z \approx 600\%$ for the particular case of relatively "thick" $Co_{67}Fe_{3.85}Ni_{1.45}B_{11.5}Si_{14.5}Mo_{1.7}$ microwire with thin glass coating (see Figure 58)[150,161]. Generally, as has been predicted, the $Co_{67}Fe_{3.85}Ni_{1.45}B_{11.5}Si_{14.5}Mo_{1.7}$ sample geometry strongly affects to the dependence of the GMI ratio on DC axial magnetic field. Such as can be observed from Figure 58, the field H_m, corresponding to the maximum of the GMI ratio, increases and $(\Delta Z/Z)_{max}$ decreases with the ρ–ratio [160].

It means that the internal stresses, arising from the difference between the thermal expansion coefficients of the metallic core and glass coating, strongly affect to the aforementioned properties.

In fact $Co_{67}Fe_{3.85}Ni_{1.45}B_{11.5}Si_{14.5}Mo_{1.7}$ composition has been selected on the basis of previous knowledge on formation of soft magnetic properties in Co- based amorphous ribbons [190].

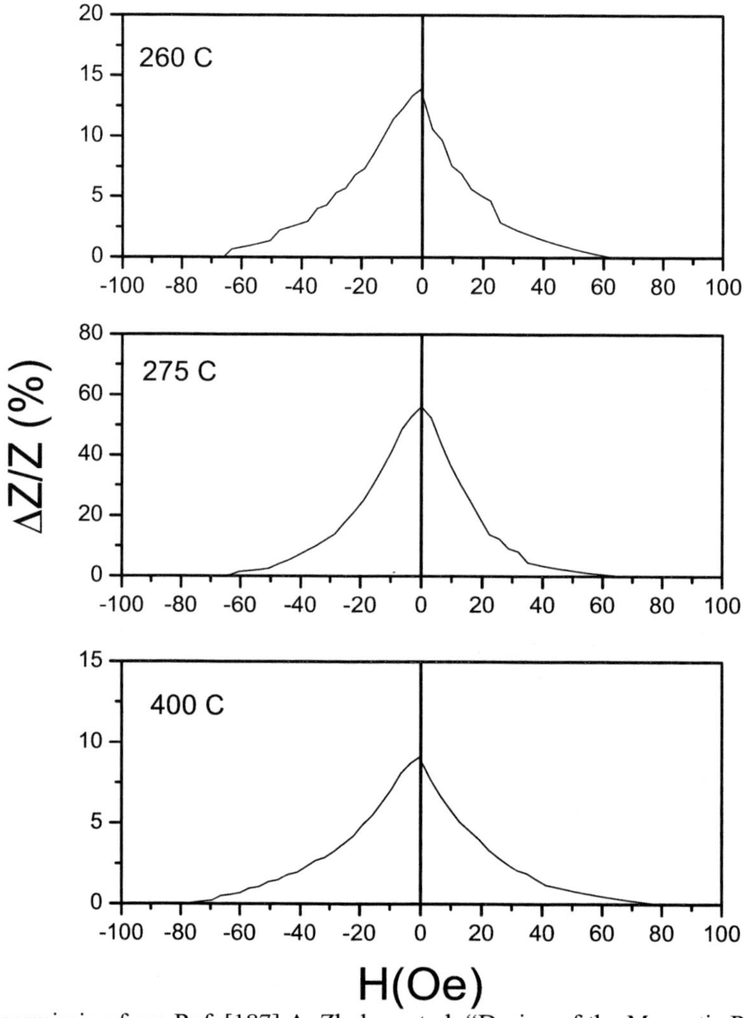

Reprinted with permission from Ref. [187] A. Zhukov et al. "Design of the Magnetic Properties of Fe-Rich, Glass-Coated Microwires for Technical Applications" Adv. Funct. Mater. 2006, 16, 675–680. Copyright WILEY-VCH (2006) (figure 4).

Figure 53. GMI effect of stress annealed of $Fe_{74}B_{13}Si_{11}C_2$ glass-coated microwire subjected to stress annealing for 0.5 hour at different temperatures.

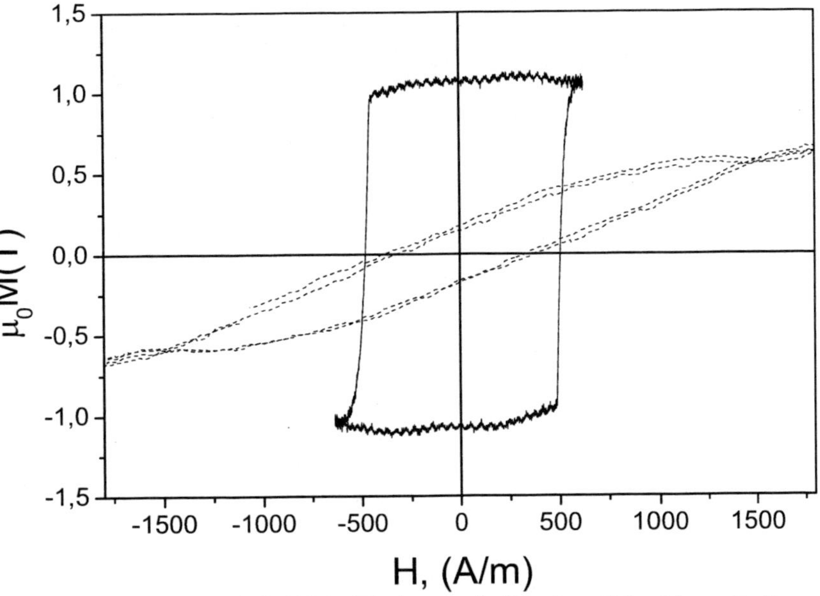

Reprinted with permission from Ref. [187] A. Zhukov et al. "Design of the Magnetic Properties of Fe-Rich, Glass-Coated Microwires for Technical Applications" Adv. Funct. Mater. 2006, 16, 675–680. Copyright WILEY-VCH (2006) (figure 5).

Figure 54. Hysteresis loop of stress annealed $Fe_{74}B_{13}Si_{11}C_2$ glass-coated microwire measured under applied stress of 500 MPA. Hysteresis loop of $Fe_{74}B_{13}Si_{11}C_2$ subjected to SA without stress has been shown for the comparison by dashed line.

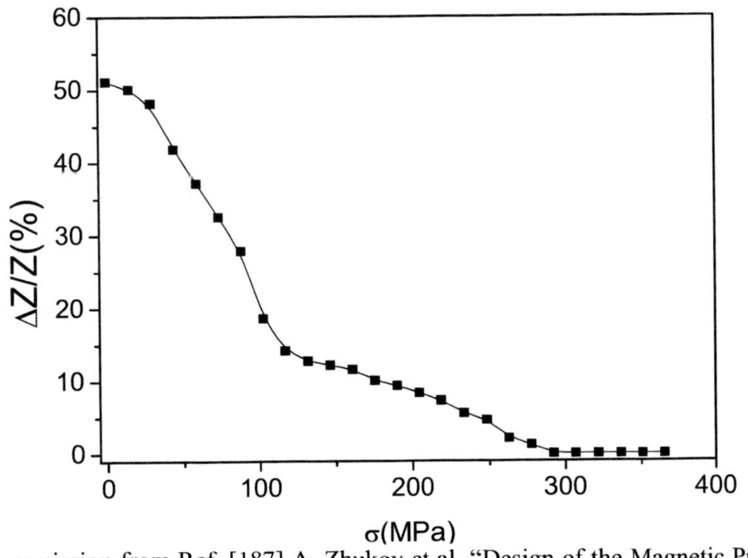

Reprinted with permission from Ref. [187] A. Zhukov et al. "Design of the Magnetic Properties of Fe-Rich, Glass-Coated Microwires for Technical Applications" Adv. Funct. Mater. 2006, 16, 675–680. Copyright WILEY-VCH (2006) (figure6).

Figure 55. Stress impedance effect of stress annealed $Fe_{74}B_{13}Si_{11}C_2$ glass-coated microwire under stress (468 MPa) at 275°C for 0.5h measured at 10 MHz for the driving current amplitude of 2 mA.

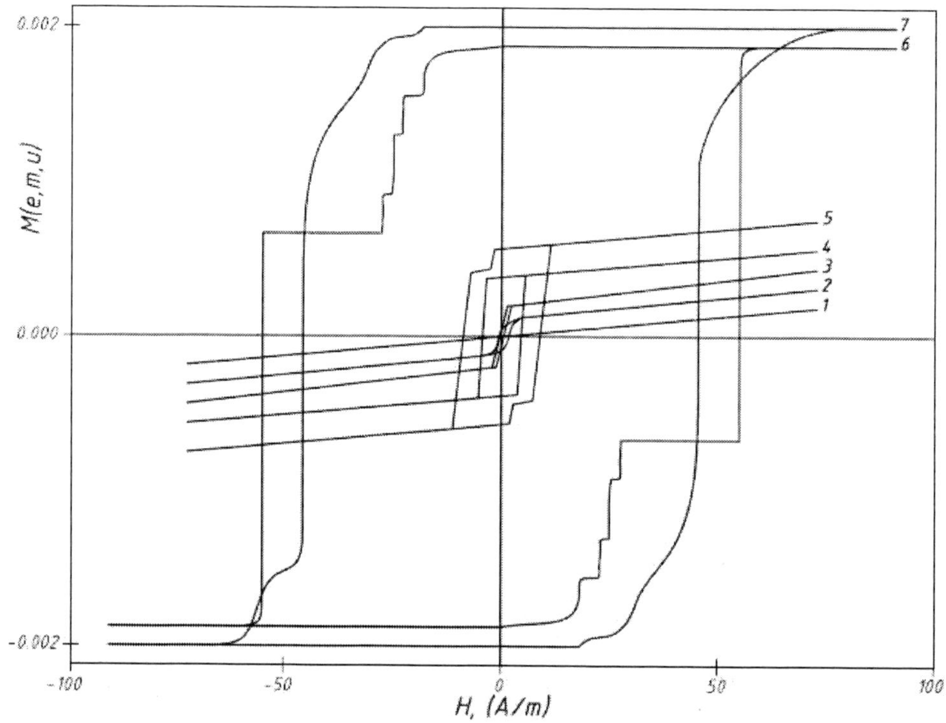

Reprinted with permission from Ref. [189] V. Larin and A. Zhukov, "Magnetic properties of microwires with amorphous structure after of thermo mechanical treatment" Phys. Stat. Sol. (a) 6, No 4, 958-961(2009). Copyright WILEY-VCH (2009). (Figure 9d).

Figure 56. Transformation of the hysteresis loop under the effect of increasing tensile stress.

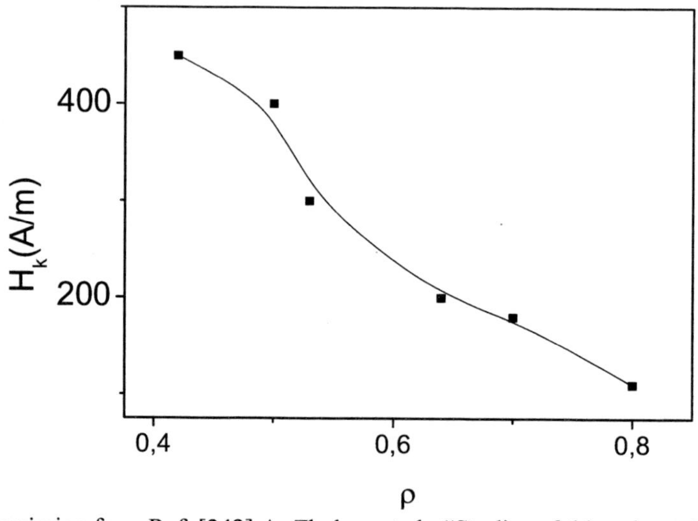

Reprinted with permission from Ref. [242] A. Zhukov et al., "Studies of thin microwires with enhanced magnetic softness and GMI effect" Phys. Status Solidi A, 206, No.4, 674-678 (2009). Copyright WILEY-VCH (2009) (Figure 3).

Figure 57. Effect of sample geometry (ρ– ratio) on magnetic anisotropy field, H_k.

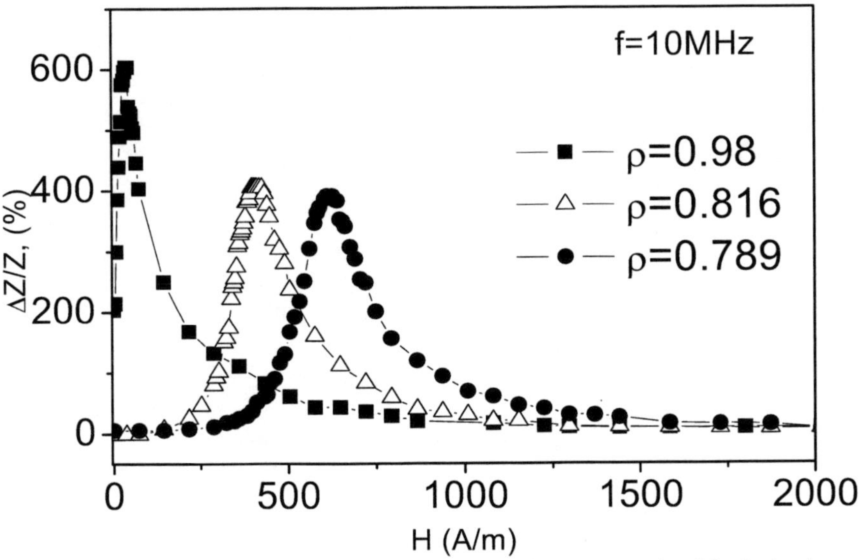

Reprinted with permission from Ref. [160] V. Zhukova et al. "Optimization of Giant Magnetoimpedance in Co-Rich Amorphous Microwires" IEEE Trans. Magn., V 38, 5 part I, 2002 pp. 3090-3092. Copyright IEEE Magnetic Society (2002) (figure 4).

Figure 58. Axial field dependence of $\Delta Z/Z$ measured at f=10 MHz and I = 0.75 mA in microwire $Co_{67}Fe_{3.85}Ni_{1.45}B_{11.5}Si_{14.5}Mo_{1.7}$ with ρ as a parameter.

It is expected that since sample geometry strongly affects the hysteretic properties, the GMI effect should be also affected by the ρ–ratio. In refs. 4, 57 magnetic field, H, dependence of real part, Z of the longitudinal wire impedance Z ($Z = Z_1+iZ_2$) has been measured. Indeed we observe (see Figure 59) that the sample geometry strongly affects the DC axial magnetic field dependence of the real part of GMI, Z_1 of $Co_{67}Fe_{3.85}Ni_{1.45}B_{11.5}Si_{14.5}Mo_{1.7}$ microwires.

Since it is known that the strength of the internal stresses is determined by the relation between metallic nucleus, d and total diameter, D, i.e. by the ρ-ratio the relaxation of such internal stresses by means of thermal treatment should drastically change both, soft magnetic behaviour and $\Delta Z/Z(H)$ dependence. This challenge is of a special importance since the fabrication of the microwires with the glass coating thickness of about 0.2 μm (sample with ρ–ratio 0,98) exhibiting very large (600%) GMI ratios are a difficult task. This is why an identification of the processing procedure, which permits to improve the GMI effect of the glass-coated microwires, is very important from the viewpoint of applications. Studies of the effect of Joule heat treatment on GMI effect have been performed for $Co_{67}Fe_{3.85}Ni_{1.45}B_{11.5}Si_{14.5}Mo_{1.7}$ microwires in a number of papers [4, 57, 160, 184]

As expected, the performed Joule heating strongly affects to the $\Delta Z/Z\ (H)$ dependence. Figure 60 shows this dependence measured for as-prepared and annealed $Co_{67}Fe_{3.85}Ni_{1.45}B_{11.5}Si_{14.5}Mo_{1.7}$ (CA treatment) samples with the frequency, f, as a parameter. It is well recognized that the maximum value of the GMI ratio, $\Delta Z/Z_{max}$, increases with f as well as after CA treatment. Besides, the value of the axial DC-field, H_m, corresponding to the maximum of the GMI ratio, increases also as a result of the Joule heating.

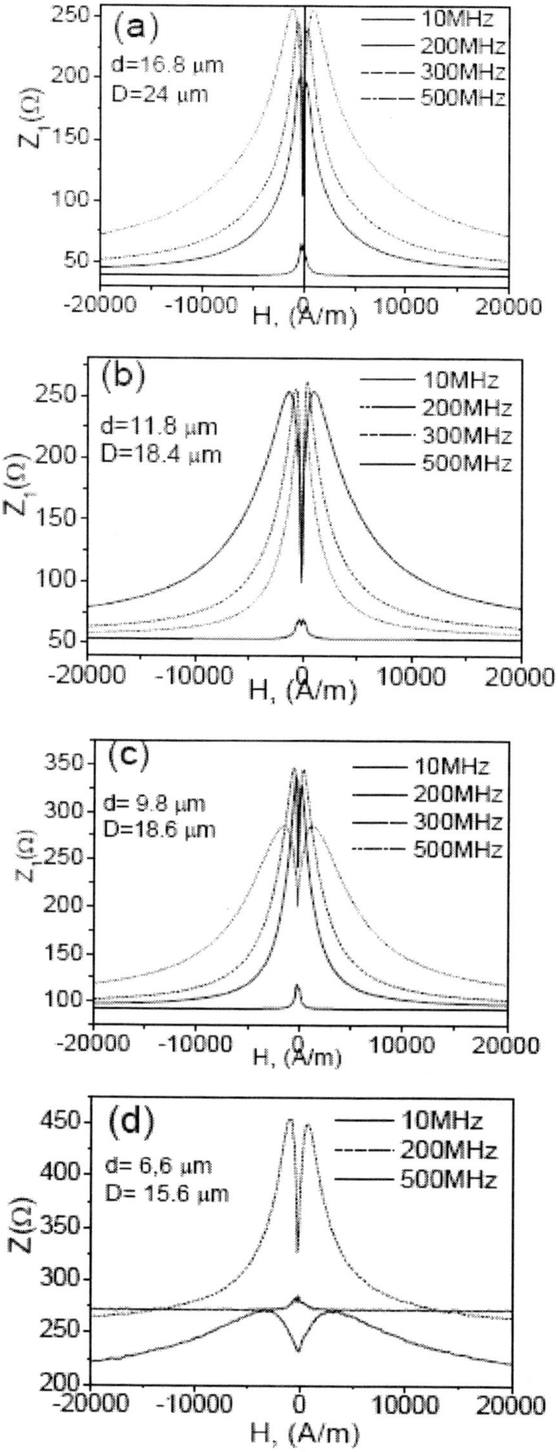

Reprinted with permission from Ref. […], A. Zhukov "Recent advances in studies of magnetically soft amorphous microwires" J. Magn. and Magn., Mater. (2009) doi:10.1016/j.jmmm.2008.11.068. Copyright Elsevier (2008) (figure 4).

Figure 59. Z (H) dependence $Co_{67}Fe_{3.85}Ni_{1.45}B_{11.5}Si_{14.5}Mo_{1.7}$ microwires with different geometry.

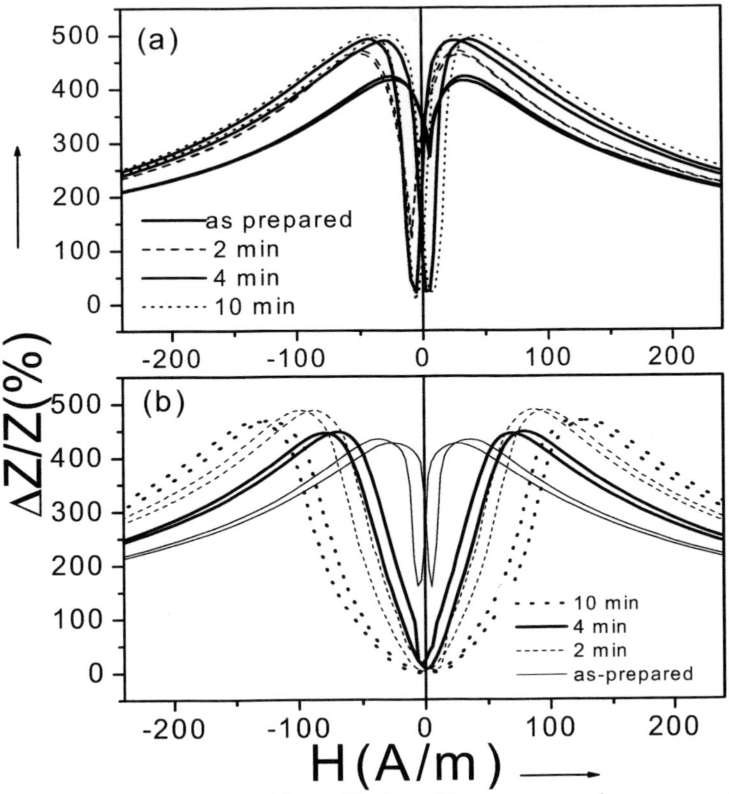

Reprinted with permission from Ref. [161], A. Zhukov "Recent research on magnetic properties of glass-coated microwires" J. Magn. and Magn., Mater. 294 (2005) 182-192. Copyright Elsevier (2005) (figure 8).

Figure 60. *ΔZ/Z(H)* dependences measured at f=30 MHz and I=1 mA in microwire subjected to CA annealing at 30 mA (a) and at 40 mA (b).

In order to illustrate better the only effect of the Joule heating, the *ΔZ/Z(H)* dependencies have been measured at the same conditions, i.e. at the fixed frequency and driving current amplitude, varying solely the Joule heating conditions. These dependencies are presented in Figure 60a (annealing current 30 mA) and in Figure 60b (40 mA).

Taking into account that CA and MFA treatments induce significant changes in the hysteresis loops of $Co_{67}Fe_{3.85}Ni_{1.45}B_{11.5}Si_{14.5}Mo_{1.7}$ microwires (see Figure 61), considerable changes of the GMI ratio dependence on the axial magnetic field have been observed. Thus, both $\Delta Z/Z_{max}$, and H_m after MFA treatment change by the different way owing to such induced magnetic anisotropy (see Figure 61).

4.3.5. Development of Extremely Thin Microwires with GMI Effect. Off-diagonal GMI.

The demagnetizing factor affects domain structure and magnetic properties of microwires when the geometric ratio of microwire diameter to its length becomes low enough [2-4, 13, 57]. Therefore, the development of miniaturized sensors requires the microwires with thinner diameter but keeping their good soft magnetic properties and GMI effect. Consequently, the development of magnetically soft microwires with reduced diameter (below 10 μm) is quite important for micro-miniaturized sensor application [7,161]. Recent studies allow improving of soft magnetic properties and GMI effect of glass coated microwires with metallic nucleus diameters generally above 10 μm [4,57] (see Figs, 7,59). Consequently, enhanced soft magnetic properties (coercivity below 10 A/m) and considerable GMI effect observed in thin microwires with metallic nucleus diameter, d, below 10 μm should be considered as significant achievement in development of materials for micro-sensor applications.

In fact one should take into account that both magnetic permeability as well as magneto-impedance has tensor character. Recently it was found that off-diagonal components possess the asymmetrical dependence on magnetic field, the necessary condition for determination the magnetic field direction [174, 175, 191, 192]. Figure 62 shows field dependence of the off-diagonal voltage response, V_{out} measured as described elsewhere [174, 175, 191, 192] in Co67.1Fe3.8Ni1.4Si14.5B11.5Mo1.7 ($\lambda_s \approx 3 \cdot 10\text{-}7$) microwire with different geometry: metallic nucleus diameter and total diameter with glass coating are 6.0/10.2, 7.0/11.0 and 8.2/13.7 μm.

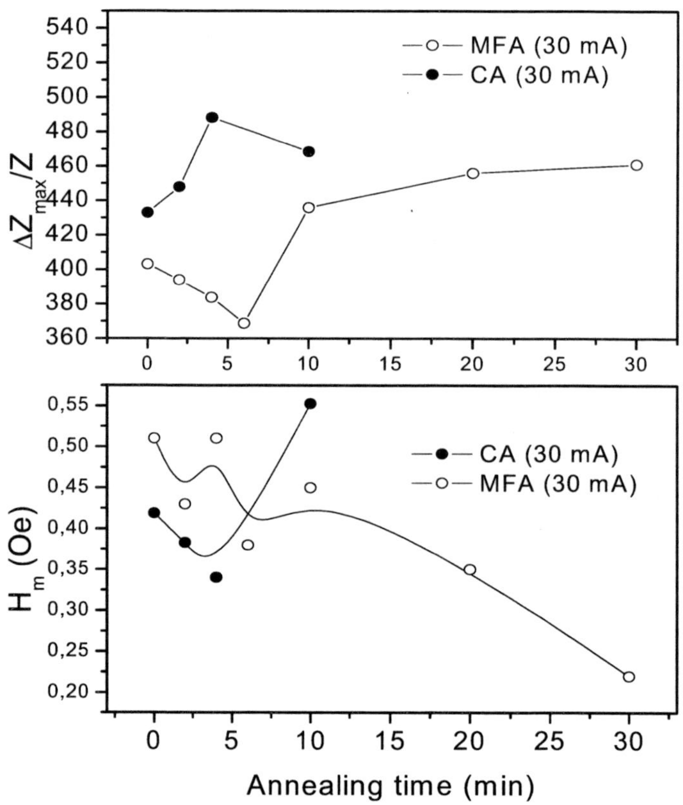

Reprinted with permission from Ref. [161], A. Zhukov "Recent research on magnetic properties of glass-coated microwires" J. Magn. and Magn., Mater. 294 (2005) 182-192. Copyright Elsevier (2005) (figure 11).

Figure 61. Effect of CA and MFA treatments on $\Delta Z/Z_{max}$ and H_m of studied microwires measured at 30 MHz and 1 mA.

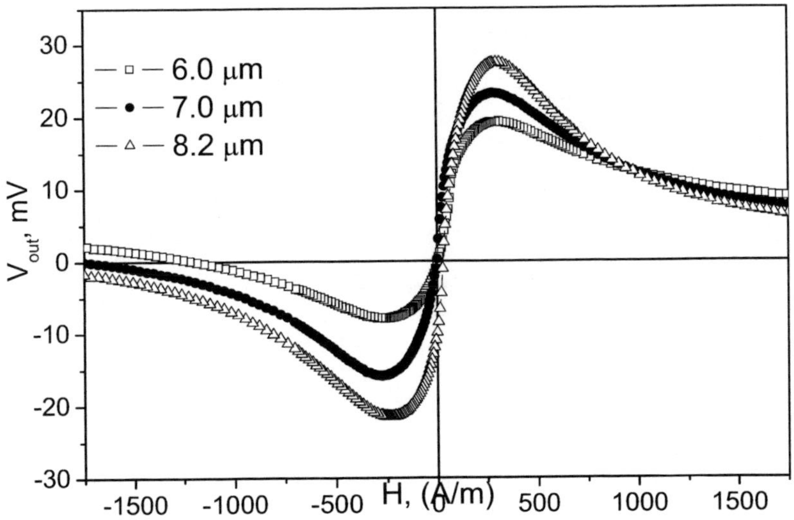

Reprinted with permission from Ref. [57], M. Ipatov et al., "Off-diagonal magneto-impedance in amorphous microwires with diameter 6–10 μm and application to linear magnetic sensors" Phys. stat. sol. (a) 205, No. 8, 1779–1782 (2008). Copyright WILEY-VCH (2008) (figure 3).

Figure 62. V_{out}(H) response of $Co_{67,1}Fe_{3,8}Ni_{1,4}$ $Si_{14,5}B_{11,5}Mo_{1,7}$ microwires with different diameter.

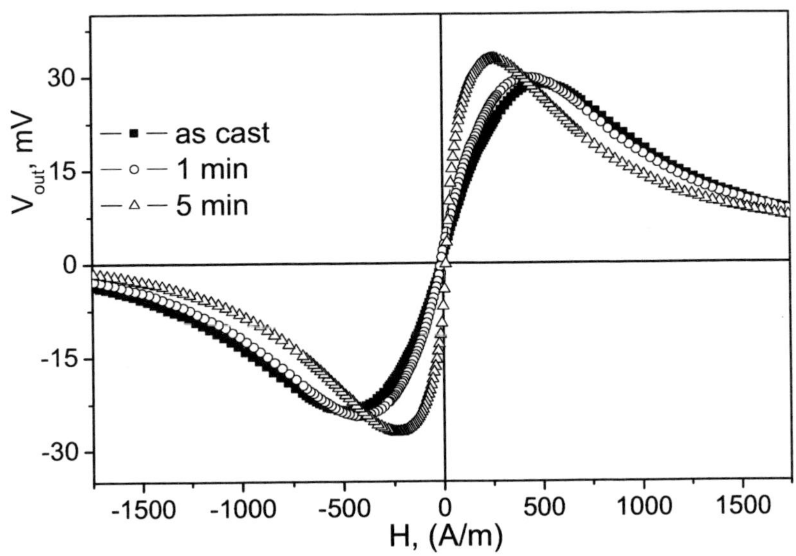

Reprinted with permission from Ref. [57], M. Ipatov et al., "Off-diagonal magneto-impedance in amorphous microwires with diameter 6–10 μm and application to linear magnetic sensors" Phys. stat. sol. (a) 205, No. 8, 1779–1782 (2008). Copyright ILEY-VCH (2008) (figure 5).

Figure 63. $V_{out}(H)$ of Joule-heated $Co_{67}Fe_{3.85}Ni_{1.45}$ $B_{11.5}Si_{14.5}Mo_{1.7}$ microwire annealed with 50 mA currents for different time.

It should be noted from Figure 63 that the $V_{out}(H)$ curves have asymmetrical shape exhibiting close to linear growth within the field range from —H_m to H_m. The H_m limits the working range of MI sensor to 240 A/m and should be associated with the anisotropy field.

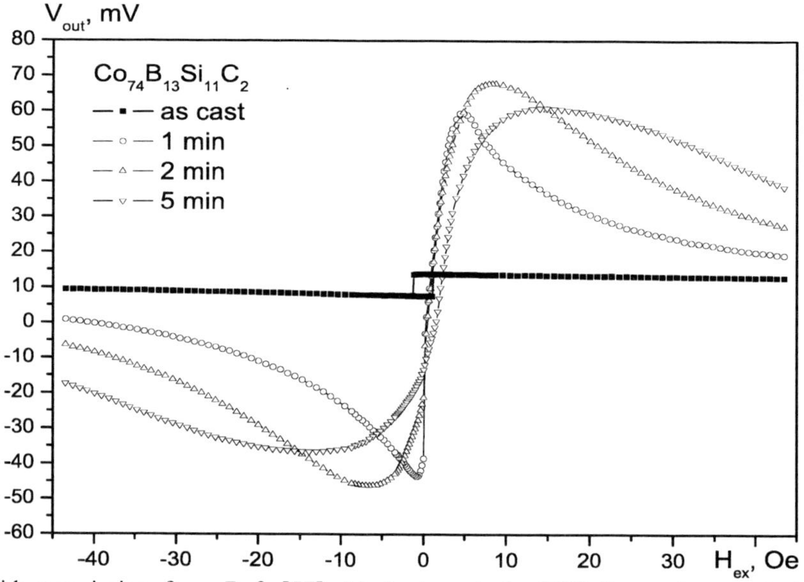

Reprinted with permission from Ref. [57], M. Ipatov et al., "Off-diagonal magneto-impedance in amorphous microwires with diameter 6–10 μm and application to linear magnetic sensors" Phys. stat. sol. (a) 205, No. 8, 1779–1782 (2008). Copyright ILEY-VCH (2008) (figure 4).

Figure 64. Field dependence of the off-diagonal voltage response of negative magnetostriction Joule-heated microwire annealed with 50 mA currents for different time.

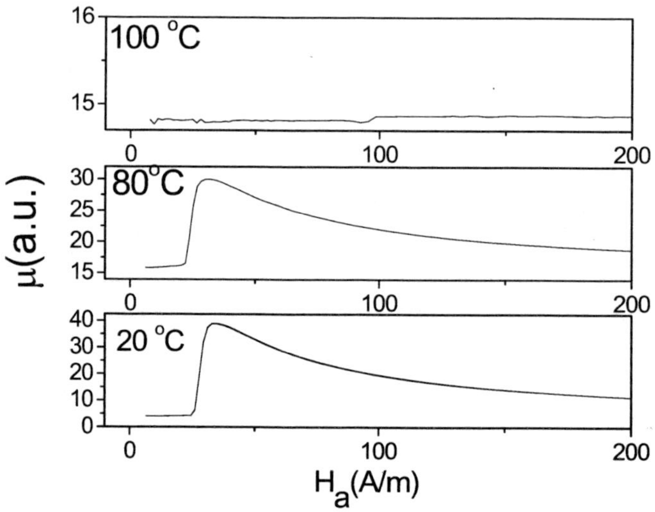

Reprinted with permission from Ref. [193], V. Zhukova et al. "Development of thin microwires with low Curie temperature for temperature sensors applications" Sensors and Actuators B 126 (2007) 318–323 Copyright Elsevier (2007) (figure 3).

Figure 65. Magnetic amplitude dependence of magnetic permeability, $\mu(H_a)$ measured in $Co_{23.67}Fe_{7.14}Ni_{43.08}B_{13.85}Si_{12.26}$ at different temperatures.

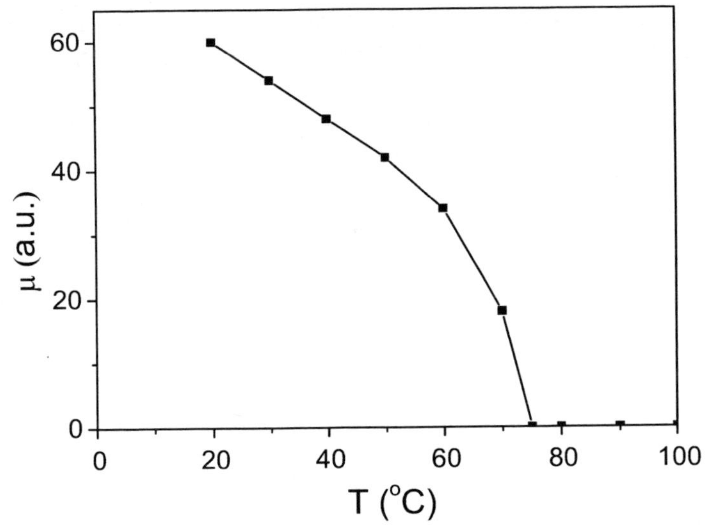

Reprinted with permission from Ref. [193], V. Zhukova et al. "Development of thin microwires with low Curie temperature for temperature sensors applications" Sensors and Actuators B 126 (2007) 318–323 Copyright Elsevier (2007) (figure 4).

Figure 66. Temperature dependence of magnetic permeability, μ (H_a) measured in $Co_{60,51}Fe_{3,99}Cr_{12,13}B_{13,53}Si_{9,84}$ microwire.

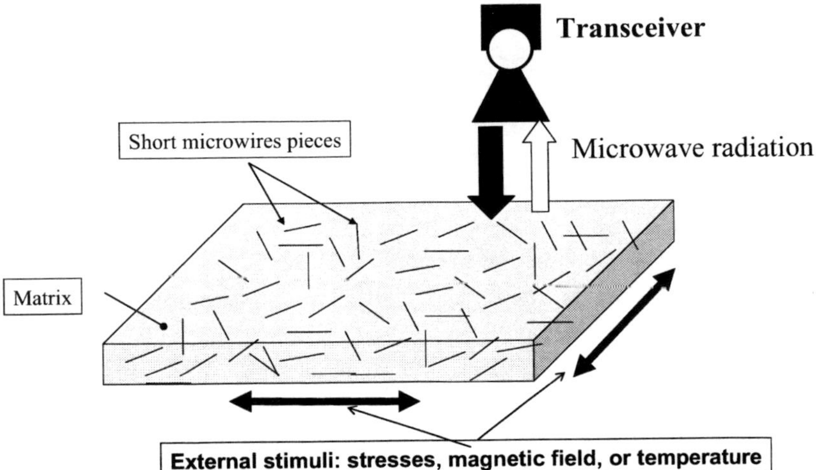

Reprinted with permission from Ref. [194] A. Zhukov et al. "Development of Stress and Temperature Sensitive Microwires for the Sensor Applications and Tuneable Composite Materials" Advances in Science and Technology Vol. 54 (2008) pp 180-186. Copyright Trans Tech Publications, Switzerland (2008) (figure 10).

Figure 67. Free-space microwave sensing technique using embedded short ferromagnetic microwires.

The influence of Joule heating on off-diagonal field characteristic of nearly zero magnetostriction $Co_{67.1}Fe_{3.8}Ni_{1.4}Si_{14.5}B_{11.5}Mo_{1.7}$ microwire with diameters 9.4/17.0 μm is shown in Figure 62. One can see that the thermal annealing with 50 mA DC current reduces the H_m from 480 A/m in as-cast state to 240 A/m after 5 min annealing.

The off-diagonal MI dependence on magnetic field in as-prepared $Co_{74}B_{13}Si_{11}C_2$ with negative magnetostriction ($\lambda_s \approx -1 \cdot 10^{-6}$) microwire with diameters 10.0/16.2 μm shows a weak negative slope with hysteresis at 1.3 Oe (Figure 64). The rather low sensitivity of negative magnetostriction microwires has been attributed to high enough internal stresses, induced in the metallic core by the glass coating. The magnetoelastic coupling of the internal stresses and large magnetostriction substantially reduce MI effect. After Joule heating significant improvement of the V_{out} signal has been observed. Observed strong dependence of $V_{out}(H)$ curves on Joule heating has been attributed mostly to decrease of magneto-elastic energy due to stress relaxation [4,57].

4.3.6. Development of Temperature Sensitive Microwires with GMI Effect

Additions of Ni and Cr in the alloys make it possible to substantially lower the Curie temperature and develop the microwires suitable for temperature detection in the range of moderate temperatures (around room temperature) [193]. For this application, wires with axial anisotropy and rectangular hysteresis would be of interest. Magnetic field dependences of the differential permeability at different temperatures as shown in Figure 65. The shape of this behavior changes with the temperature and the permeability drops abruptly at the approach to the Curie temperature. Figure 66 shows the maximum of the permeability vs. temperature for $Co_{60,51}Fe_{3,99}Cr_{12,13}B_{13,53}Si_{9,84}$ microwire having a narrow rectangular loop at room temperature. In this case, the Curie temperature is found to be about 75 °C.

The developed microwires are proposed for completely new applications in sensory materials for non-destructive remote control of stress, strain and temperature [9, 174, 194-195]. These composites may contain short pieces of microwires embedded into a polymeric matrix. The wire impedance sensitive to its specially designed magnetic structure will be responsible for producing a controlled microwave dielectric response. A schematics of such composite and microwave interrogation is shown in Figure 67. This research is at the initial stage of its developments.

Resuming, the composite origin and the additional magnetoelastic energy related with the presence of the glass coating require more attention for tailoring of magnetic properties of microwires. Post fabrication processing, i.e. special thermal treatment, precise control of their geometry, chemical composition of the metallic nucleus etc. From the point of view of magnetic properties, excellent magnetic softness can be achieved in these glass-coated microwires after rigorous control of their parameters (geometry, chemical composition etc) and using post fabrication processing of such materials.

Thin amorphous wires with excellent magnetic softness and large magnetoimpedance and stress impedance effects have been developed. It is demonstrated that the magnetic properties can be tailored by an appropriate selection of the alloy composition, wire geometry and thermal treatments. Stress annealing of Fe-rich glass-coated microwire makes it possible to considerably change the effective anisotropy and magnetic configuration. By varying the annealing time and temperature, an optimal magnetic configuration in Fe-rich microwires is achieved in terms of magnetic anisotropy, magnetic softness, GMI and SI. It is demonstrated that additions of Ni or Cr results in decreasing the Curie temperature and high temperature sensitivity of the magnetisation and permeability in the corresponding temperature region. It is foreseen, that the developed microwires may find applications in stress/temperature sensors, as well as components for smart sensory materials.

4.4. Other Ways of Tailoring of Magnetic Properties

From aforementioned description it has been assumed that magnetoelastic anisotropy and shape magnetic ansitropy are main factors determining hysteretic magnetic properties and GMI behaviour of the glass coated microwires. Therefore certain efforts have been paid to control the parameters responsible for both magnetoelastic and shape anisotropy, such as:

a) Metallic nucleus chemical composition
b) Samples geometry
c) Thermal treatments

On the other hand, certain effors have been paid to tailoring of magnetic properties through creation of artificial magnetic structure using partial devitrification and/or crystallization of the metallic nucleus and/or creating microwires arrays using microwires of different or even of the same composition and/or magnetic anisotropy either changing of the characteristics and even the material of glass coating.

Here we'll present overview of the existing knowledge on tailoring of magnetic properties using these methods

4.4.1. Interaction between Microwires

Simplest way to tailor magnetic properties of single microwire is to use the effect of the dipolar interaction among magnetic wires [196-199] showing that when two or more wires are placed forming a linear array, their hysteresis loops present plateaux and shifts caused by the effect of the dipolar field generated by the neighboring wires.

The results of SQUID magnetometer measurements in $Fe_{65}Si_{15}B_{15}C_5$ microwires are presented in Figure 68. The distance between the wires was equal to the double thickness of glass coating (7.4 μm). The hysteresis loop for one wire (Figure 68a) has a square shape that is associated with a single large Barkhausen jump for the magnetization process. The increase of the number of the wires causes the increase of the number of jumps (Figs 68 b-d). The magnetic field of the first jump decreases (and changes the sign), when the magnetic field of the last jump increases. The value of the magnetic field of the last jump for ten wires (Figure 68d) exceeds five times the magnetic field of the jump for one wire. This effect is connected to the mutual interaction in the multi-wire system [196-199]. As a consequence of the stray fields, each one of the wires could be affected by the additional magnetic field. The superposition of the external and the stray fields evokes the magnetization reversal in one of the samples when the external field is less than in the case of a single wire (see schematic picture in Figure 69). The successive magnetization reversal in the wires changes the magnetic distribution in the multi-wire system and the jumps take place in the superposition of the stray fields of the remagnetized and the non-remagnetized wires. The shape of the hysteresis loop for the ten-wire system (Figure 68d) testifies that each wire in this system is influenced by the stray fields of all the surrounding wires. Surface hysteresis loops measured by Kerr effect (MOKE) also give similar results. Frequency dependence of the switching field has been studied using the Kerr effect technique for the two-wire system. The results of the experiments are presented in the Figure 70. H_1 and H_2 are the switching fields in the first and second wires. An increase of H_1 and H_2 is observed as frequency increases. The difference between H_1 and H_2 decreases and disappears at a frequency of about1200 Hz. Frequency dependencies of the switching field of single wires and ribbons have been analyzed using the modelling presented in Ref. [200, 201]. Observed increasing of the H_1 and H_2 with frequency is generally agreed with previously reported behaviour for different magnetic materials [200, 201].

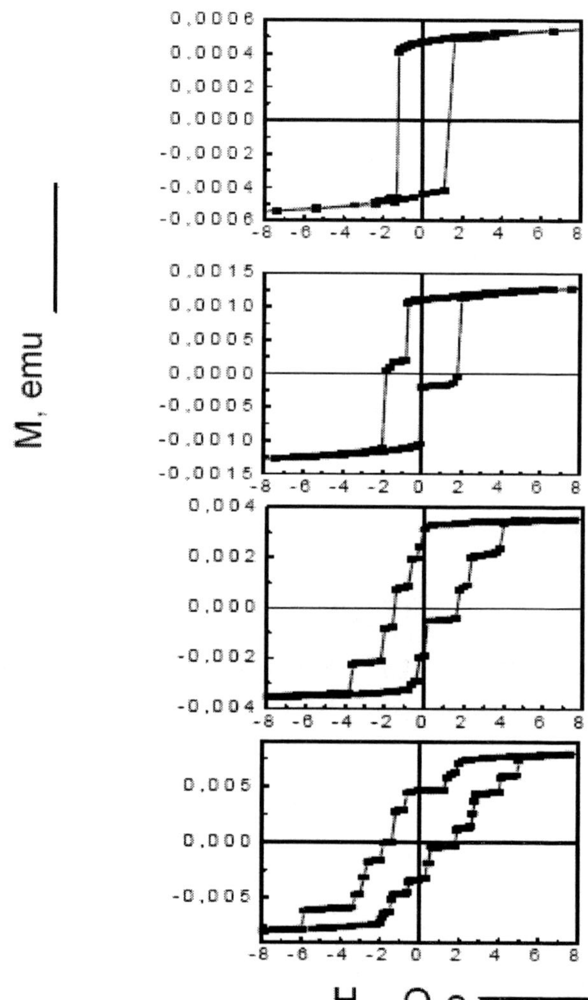

Figure 68. Hysteresis loops of $Fe_{65}B_{15}Si_{15}C_5$ microwire measured for (a) 1 microwire (b) 2 microwires (c) 5 microwires and (d) 10 microwires.

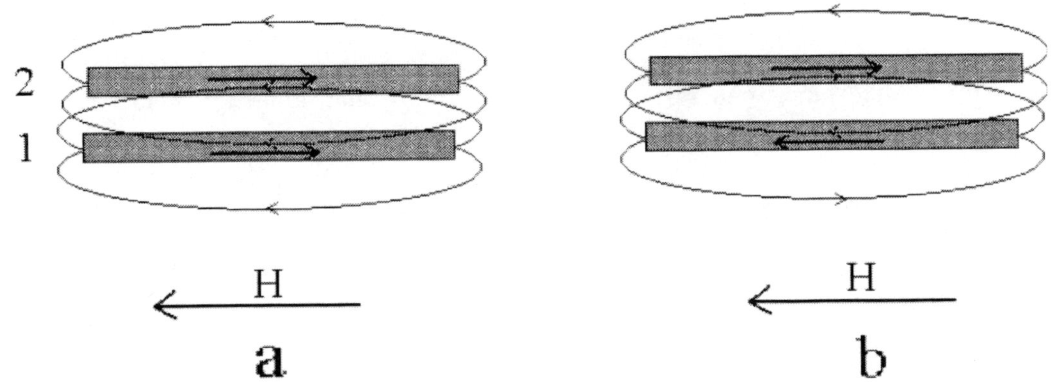

Reprinted with permission from Ref. [196], A.Chizhik et al., "Surface and volume hysteresis loops of Fe-rich glass-coated microwires" J. of Non-Cryst. Solids 287 (2001) 374-379. Copyright Elsevier (2001) (figure 3).

Figure 69. Schematic picture explaining mechanism of microwires interaction.

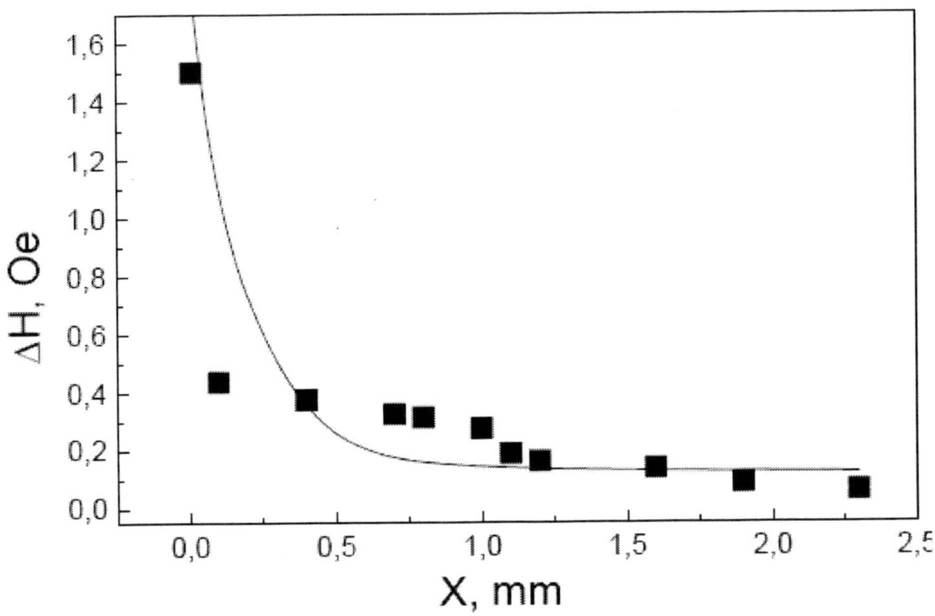

Reprinted with permission from Ref. [197], A.Chizhik et al., "Interaction between Fe-rich ferromagnetic glass-coated microwires" J. Magn. and Magn., Mater. 249 (2002) 99-103. Copyright Elsevier (2002) (figure 4).

Figure 70. Dependence of the difference between the two jumps on the distance between the wires in the two-microwire system. Frequency of the magnetic field is 50 Hz.

The decrease of the ΔH (difference between the two Barkhausen jumps) with frequency (see Figure 70) could be related to the finite value of the domain wall velocity. As was above noted the different values of the magnetic field of the two jumps could be ascribed to the influence of the stray field on the magnetization reversal. In particular, the second jump is happened in the superposition of the external field and the stray field of the wire, which is first re-magnetized. There are two time-dependent processes: the arising of the external magnetic field and the formation of the stray field during the magnetization reversal. When the values of the time of these two processes are close, the influence of the stray field decreases and accordingly ΔH decreases. The limit situation takes place at high frequency when there is no difference between the jumps. The field value necessary for the magnetization reversal in the second wire has reached considerably earlier than the end of magnetization reversal in the first wire. The distribution of stray field around the first wire is not formed yet and, therefore, there is no influence of the re-magnetized wire on the magnetization reversal in the second wire.

The dependence of ΔH differences between the two jumps for the two-wire system has been measured by the conventional fluxmetric method (Figure 71). The decrease of the ΔH differences with the distance between the wires is associated with the decrease of the stray field. The performed experiments show that a mutual influence is displayed up to a distance, X, of 2.5mm between the wires. For a distance longer than 2.5mm the magnetization reversal appears independently in the wires. The experimental results have been approximated by the dependence $\Delta H \sim X^{-3}$ (line in Figure 71), taking into account such dependence of the stray field value on the distance from the magnetized sample. This description of the experimental dates by the above dependence suggests that the influence of the stray field is the main reason of the observed successive jumps in the multi-wire system.

The problem of interaction between thin wires during magnetic measurements and even for GMI behaviour seems to be important from the point of view of applications as well as from the point of view of understanding of the magnetic phenomena. The design of more sophisticated magnetic sensors requires the use of different magnetic configurations using a few samples to detect the change of the magnetic field or when the few field sensitive elements (wires, thin films) are place close to each other. Improvement of the flux gate characteristics has been suggested when non-linearity of the hysteresis loops has been introduced [202-204]. Therefore, effect of microwire interaction on GMI effect has been studied in an array of microwires containing a number of $Co_{59.2}Fe_{4.8}Ni_{11}B_{14}Si_{11}$ amorphous glass-coated microwires exhibiting nearly-zero magnetostriction constant [202].

The GMI effect was measured in arrays of this microwire containing a number (*n*) of thin glass-coated microwires varying from 1-30. Figure 72 shows typical *ΔZ/Z(H)* curves obtained for three arrays (according the number of microwires *n* involve in the array). As it is expected, these curves are symmetric respect to the direction of the axial applied magnetic field, *H*, with a maximum value of *ΔZ/Z* ratio for a value of the applied magnetic field, H_{max}, that in some way could be ascribed to the magnetic anisotropy field. Nevertheless, the *ΔZ/Z* maximum ratio as well as the H_{max} value depends significantly on *n* parameter. In fact, the figures 73 a and b present the evolution of the GMI maximum ratio with the number of microwires (*n*) measured at 10 MHz (a) and 500 MHz (b). This GMI maximum decreases with *n* reflecting the interaction effect among the microwires of the array probably due to a shielding phenomenon of the applied magnetic field. In addition, the changes of the GMI response are larger at 500 MHz reaching *ΔZ/Z* values up to 350 %, which could interpreted considering that circular magnetic permeability, μ_ϕ, increases with the frequency since the skin penetration increases. Consequently, the circumferential anisotropy very near of the surface becomes stronger, giving rise a significant increase of μ_ϕ with the frequency and, therefore, *ΔZ/Z* changes result to be higher for 500 MHz than those obtained at 10 MHz.

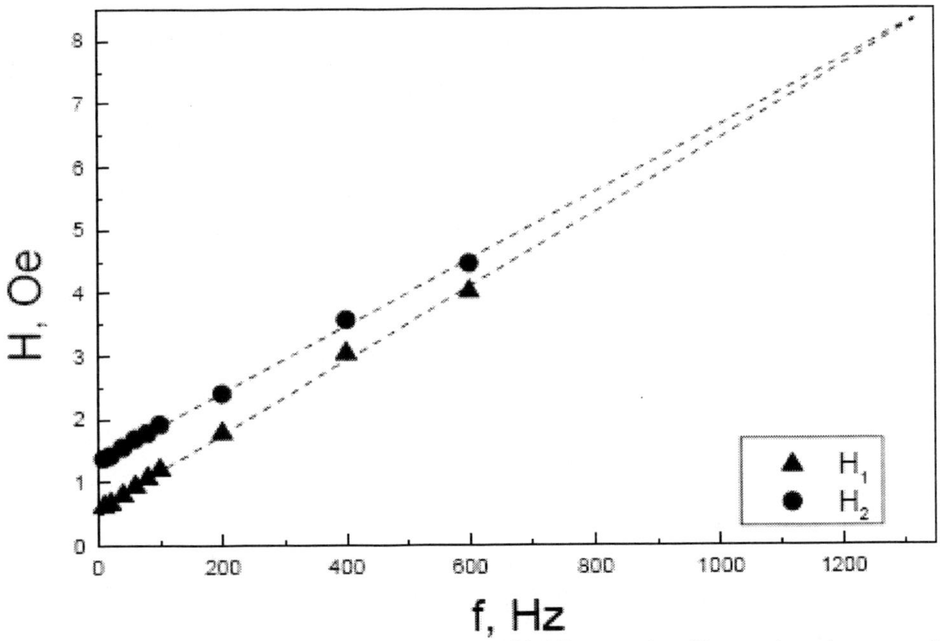

Reprinted with permission from Ref. [197], A.Chizhik et al., "Interaction between Fe-rich ferromagnetic glass-coated microwires" J. Magn. and Magn., Mater. 249 (2002) 99-103. Copyright Elsevier (2002) (figure 3).

Figure 71. Frequency Dependence of the difference between the two jumps in the two-microwire system.

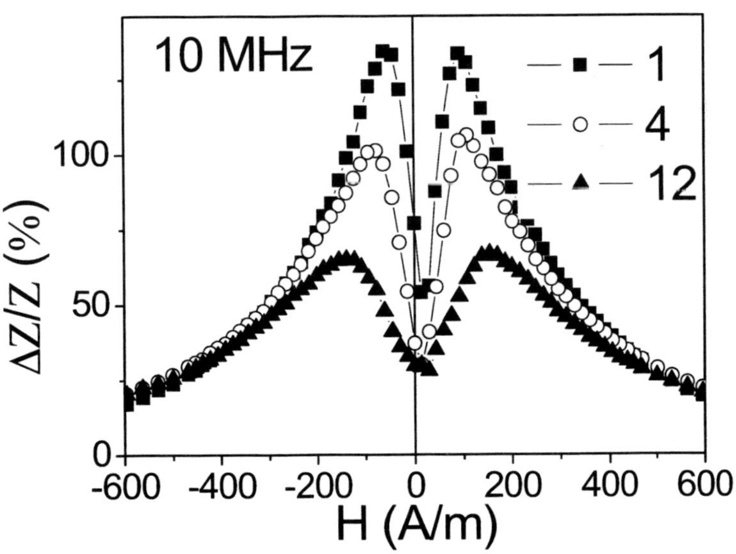

Reprinted with permission from Ref. [202], C.Garcia et al., "Effect of Interaction on GMI Effect in a System of Few Thin Wires" Sensor Letters Vol.5, 1–3, 2007. Copyright American Scientific Publishers (2007) (figure 1).

Figure 72. *ΔZ/Z(H)* curves obtained for 1, 4 and 12 microwires.

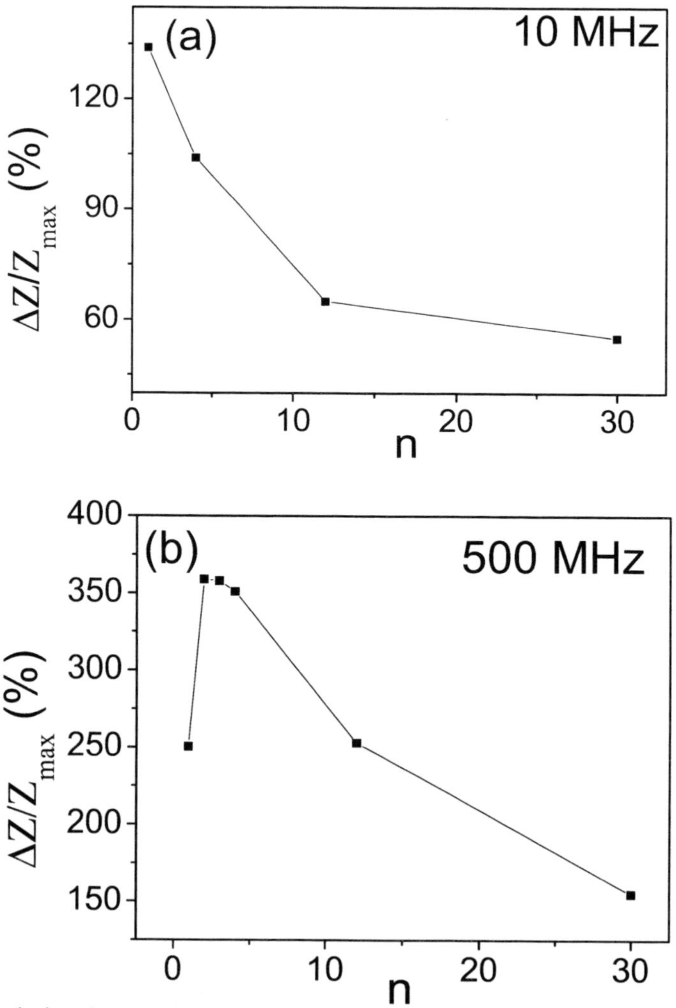

Reprinted with permission from Ref. [202], C.Garcia et al., "Effect of Interaction on GMI Effect in a System of Few Thin Wires" Sensor Letters Vol.5, 1–3, 2007. Copyright American Scientific Publishers (2007) (figure 2).

Figure 73. $\Delta Z/Z_{max}$ dependence on the number of microwires (n) measured at 10 MHz (a) and 500 MHz (b).

The figures 74 a, b show the variation of H_{max} with n parameter for the frequency of 10 (a) and 500 (b) MHz respectively. The increase of H_{max} with n parameter at 10 MHz could be related to shielding effects which drastically modify the circumferential permeability (μ_ϕ). However, it is expected very low value of μ_ϕ at 500 MHz and, consequently, such shielding effect should be less significant than at 10 MHz.

Recently the induction of non-linearity of the hysteresis loop using microwires interaction as well as an improvement of odd harmonics detected by pick-up coil has been reported when studying interaction identical (Fe-rich) as well as different (Fe-rich with rectangular hysteresis loop and Co-rich with inclined hysteresis loops) microwires [199]. Peculiar hysteresis loops with well pronounced non-linearity have been obtained by placing together $Fe_{74}B_{13}Si_{11}C_2$ microwire with coercivity about 150 A/m and

$Co_{66,94}Fe_{3,83}Ni_{1,94}B_{11,51}Si_{14.59}$ $Mo_{1,69}$ microwire with coercivity about 7 A/m (see Figure 75)[205]. Increasing the amplitude of magnetic field above coercivity of Fe-rich sample unusual hysteresis loop with well defined non-linearity has been obtained. Certainly use of microwire arrays can serve for designing of hysteretic properties for sensor applications.

Summarizing, the mutual influence of the stray field of the wires is reflected in the volume and surface hysteresis loops and on GMI effect. The transformation of the hysteresis loop with the increase of the distance between wires has been observed. The obtained results are in agreement with the model of mutual influence of the wires. These results can be used for tailoring of the hysteresis loops and GMI response of the arrays of microwires

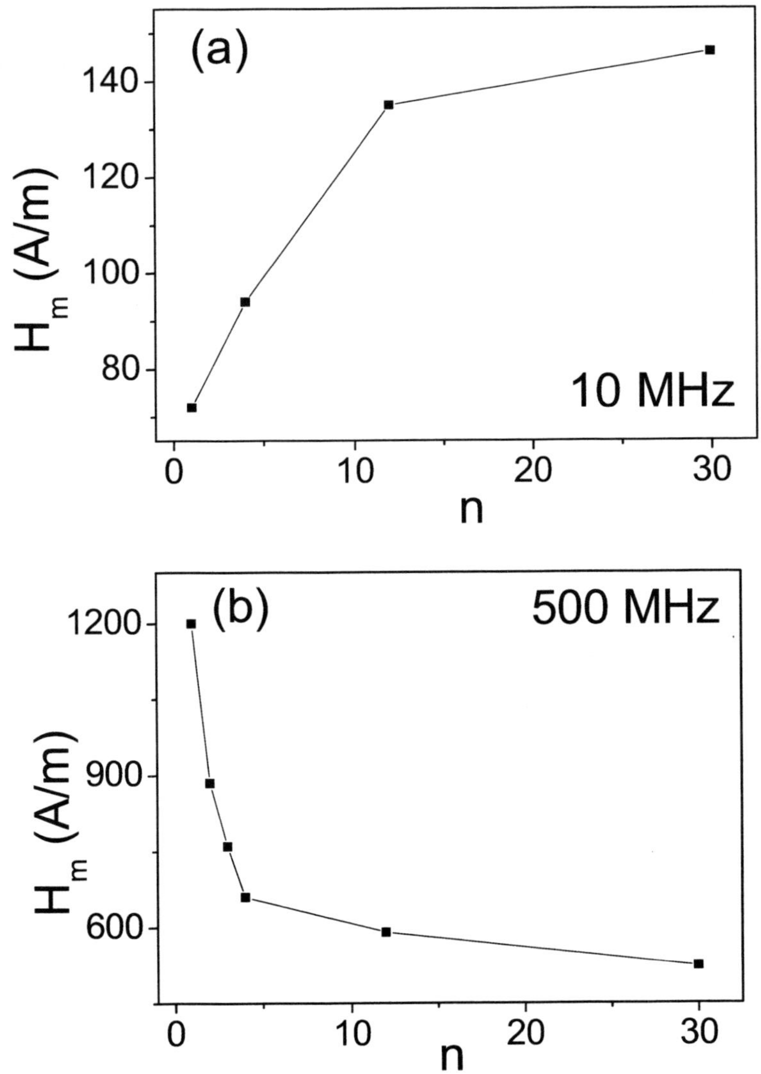

Reprinted with permission from Ref. [202], C.Garcia et al., "Effect of Interaction on GMI Effect in a System of Few Thin Wires" Sensor Letters Vol.5, 1–3, 2007. Copyright American Scientific Publishers (2007) (figure 3).

Figure 74. H_{max} dependence on the number of microwires (n) measured at 10 MHz (a) and 500 MHz (b).

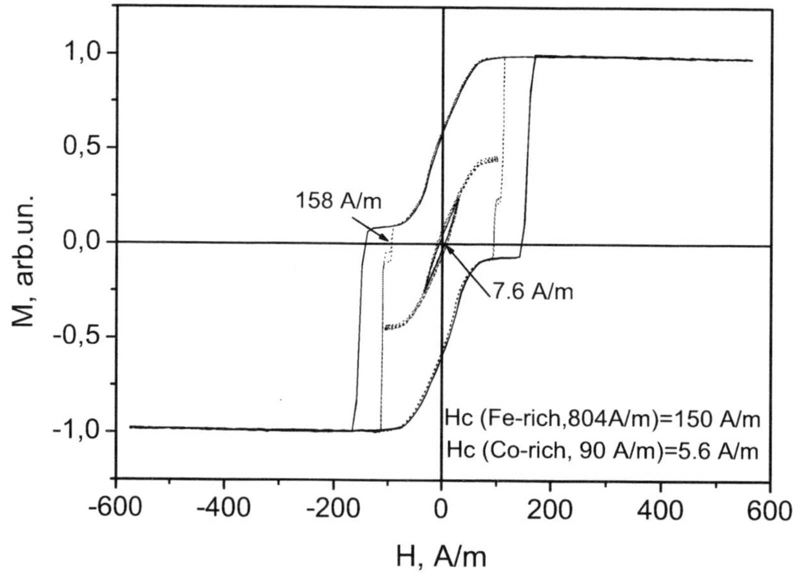

Figure 75. Hysteresis loops of $Fe_{74}B_{13}Si_{11}C_2$ and $Co_{66,94}Fe_{3,83}Ni_{1,94}B_{11,51}Si_{14.59}Mo_{1,69}$ microwires measured at different amplitude of magnetic field (from 33 till 570 A/m) V. Samsonova et al. (to be published).

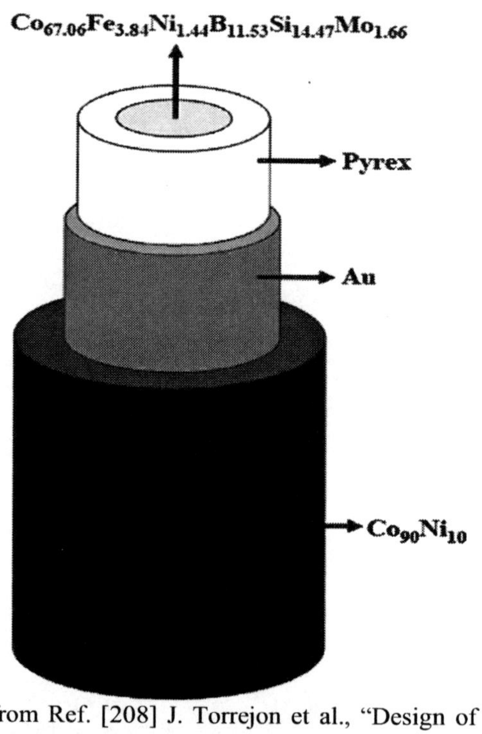

Reprinted with permission from Ref. [208] J. Torrejon et al., "Design of multilayer microwires with controllable magnetic properties: Magnetostatic and magnetoelastic coupling" Acta Materialia 55 (2007) pp.4271–4276. Copyright Elsevier V. L. (2007) (figure 1).

Figure 76. A scheme of the multilayer magnetic microwire: soft magnetic nucleus (light grey), intermediate glass layer (white), nanometric Au layer (dark grey) and hard magnetic outer shell (black).

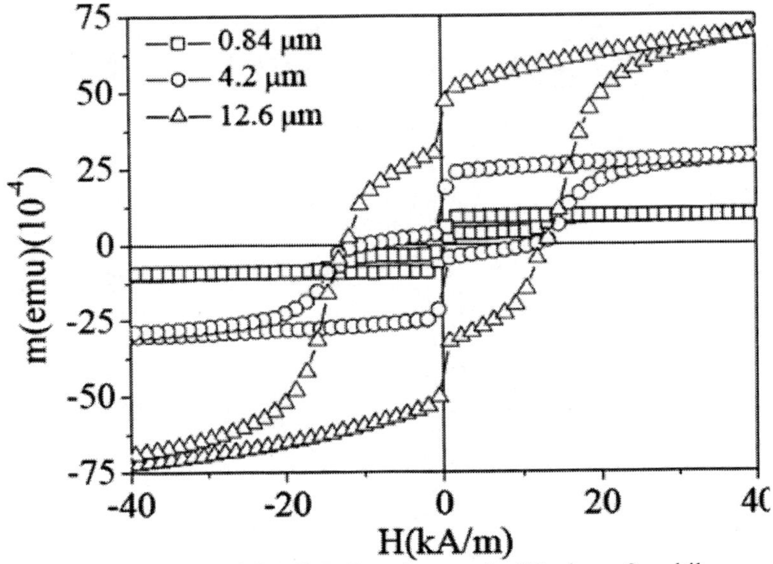

Reprinted with permission from Ref. [208] J. Torrejon et al., "Design of multilayer microwires with controllable magnetic properties: Magnetostatic and magnetoelastic coupling" Acta Materialia 55 (2007) pp.4271–4276. Copyright Elsevier V. L. (2007) (figure 2).

Figure 77. High-field hysteresis loops for soft/hard two-phase multilayermicrowire for selected thicknesses of the outer hard magnetic layer at room temperature.

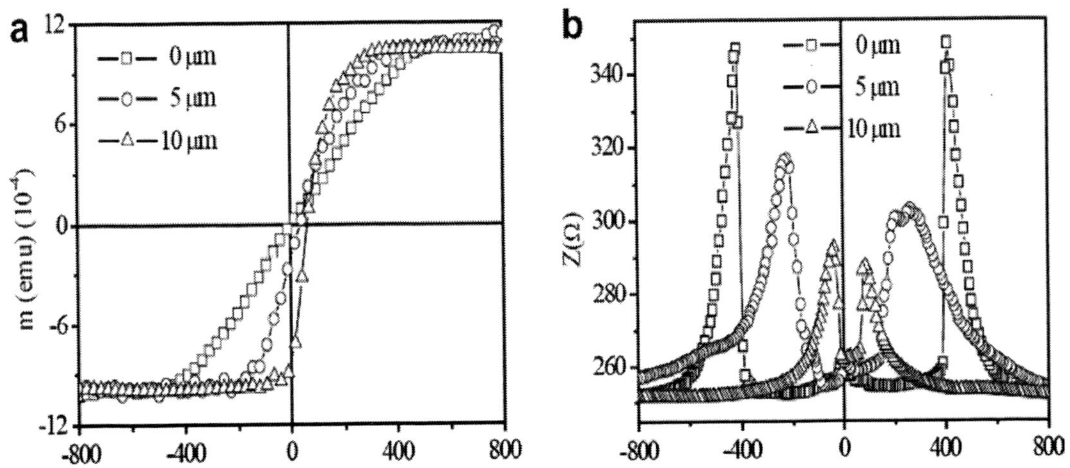

Reprinted with permission from Ref. [208] J. Torrejon et al., "Design of multilayer microwires with controllable magnetic properties: Magnetostatic and magnetoelastic coupling" Acta Materialia 55 (2007) pp.4271–4276. Copyright Elsevier V. L. (2007) (figure 7).

Figure 78. Low-field hysteresis loops (after premagnetizing at +1600 kA m^{-1}) (a) and GMI measurements (after premagnetizing at +30 kA m^{-1}) (b) for multilayer microwires with different thickness of CoNi layer.

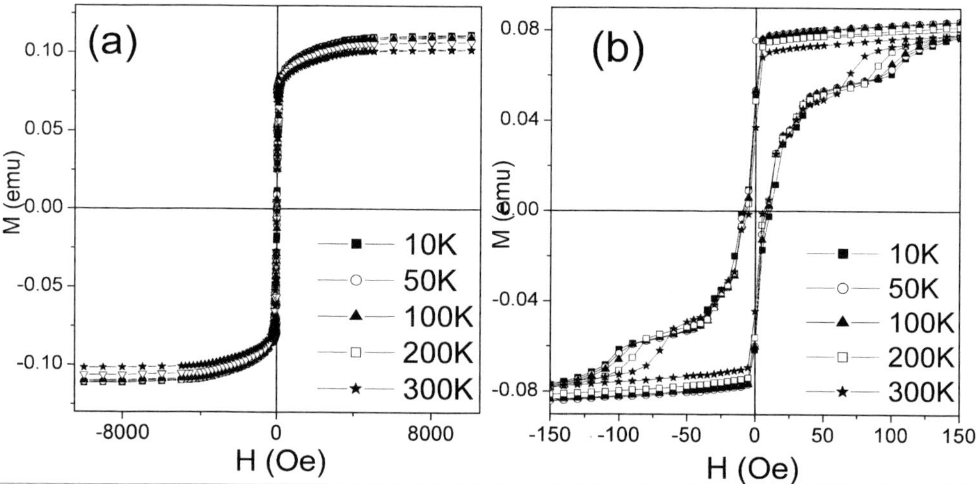

Reprinted with permission from Ref. [73] A. Zhukov et al., "Fabrication and magnetic properties of $Cu_{50}(Fe_{69}Si_{10}B_{16}C_5)_{50}$ thin microwires" Journal of Non-Crystalline Solids 353 (2007) 922–924. Copyright Elsevier V. L. (2007). (figure 1).

Figure 79. Magnetization curves of Cu50-(Fe69Si10B15C5)50 microwires in as-prepared state.

4.4.2. Multi-layered Microwires

Few years ago a new family of multilayer microwires was introduced. This novel material was obtained through the combination of different preparation techniques, including Taylor-Ulitovski method, sputtering and electroplating. The novelty of this material lays in the nature of the coupling observed between the various magnetic layers (Figure 76) [206]. As previously reposted [207], the presence of an external magnetic shell around a glass-coated microwire induces a longitudinal anisotropy in the nucleus of the latter, drastically modifying its magnetoinductive behaviour making it very sensitive to the presence of an applied external stress and temperature. A peculiar consequence of the bilayer character of the microwire is that the modification of temperature induces additional stresses just as a consequence of different thermal expansion of the layers. Because of the large differences between the thermal expansion coefficients of the glass layer and the metallic nucleus, mechanical stresses arise in the nucleus and the modified inductance, L, can be directly proportioned to the change in the external temperature.

Multilayer magnetic microwires have been prepared with suitable composition and characteristics in order to obtain optimized soft/hard two-phase material where a magnetostatic bias field can be properly detected. The precursor microwire consists of an amorphous nucleus, with nominal composition $Co_{67.06}Fe_{3.84}Ni_{1.44}B_{11.53}Si_{14.47}Mo_{1.66}$ and 17.4 μm in diameter, and vanishing magnetostriction of about 10^{-7}. It is first coated by an intermediate non-magnetic Pyrex layer, 12.1 lm thick, produced during the amorphous microwire fabrication by the quenching and drawing method [208]. An Au nanolayer, 30 nm thick, is then sputtered onto the Pyrex, which plays the role of electrode/substrate for the final electroplating of the magnetically harder CoNi outer microlayer (Figure 76).

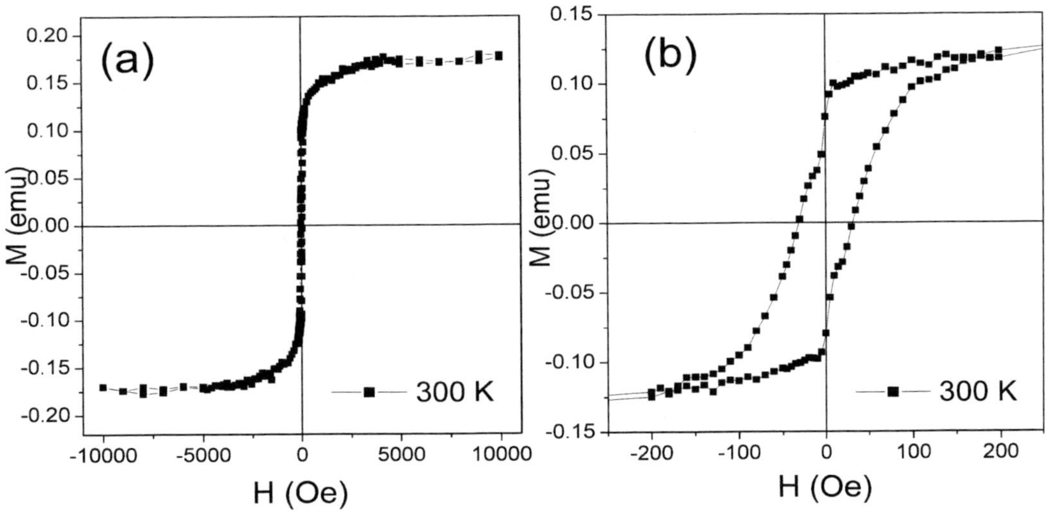

Reprinted with permission from Ref. [73] A. Zhukov et al., "Fabrication and magnetic properties of $Cu_{50}(Fe_{69}Si_{10}B_{16}C_5)_{50}$ thin microwires" Journal of Non-Crystalline Solids 353 (2007) 922–924. Copyright Elsevier V. L. (2007) (figure 4).

Figure 80. Magnetization curves of Cu_{50}-$(Fe_{69}Si_{10}B_{15}C_5)_{50}$ microwires annealed at 573K.

As observed, the magnetization process of such multilayer magnetic microwires takes place in two steps, denoting a two-magnetic phase character: a giant Barkhausen jump in very low-field, of a few A/m, that must be ascribed to the magnetic amorphous nucleus; and an additional spread jump (in the range between 10 and 20 kA/m), corresponding to the harder CoNi outer shell (Figure 77). The fractional magnetization jump and the effective coercivity of the loops depend on the thickness and composition of the hard shell [206-208].

On the other hand, the magnetoelastic properties have been also studied as a function of the CoNi thickness, t_{CoNi}. During the electroplating process, internal stresses are induced in the nucleus, affecting its domain structure. Figure 78 a and b shows low-field (±800 A/m) hysteresis loops and GMI measurements of the nucleus after premagnetizing at +1600 and +30 kA/m, respectively, for a multilayer microwire with different t_{CoNi}. In both graphics a reduction of the effective transverse anisotropy with the t_{CoNi} is observed. Thus it can be concluded that the electroplated layer generates compressive stresses in the amorphous nucleus, inducing an axial anisotropy field.

In summary, various kinds of magnetic multilayer microwires can be prepared by combined rapid solidification, sputtering and electrodeposition techniques. A multilayer microwire with a tailored magnetic response has been designed and fabricated that consists of an ultrasoft magnetic nucleus, intermediate non-magnetic layers and a semi-hard magnetic outer shell. The magnetostatic coupling between the two-magnetic-phases has been proved to result in an effective bias field, which gives rise to an effect equivalent to that of FM/AFM coupling. The strength of this bias field is controlled by the magnetic character of the outer layer and its thickness. On the other hand, magnetoelastic interactions between the nucleus and the external layers results in an effective axial or transverse (circular) anisotropy that determines the rate of magnetization reversal of the soft nucleus. Additional exchange coupling between magnetic phases must be discounted due to the presence of intermediate microlayers.

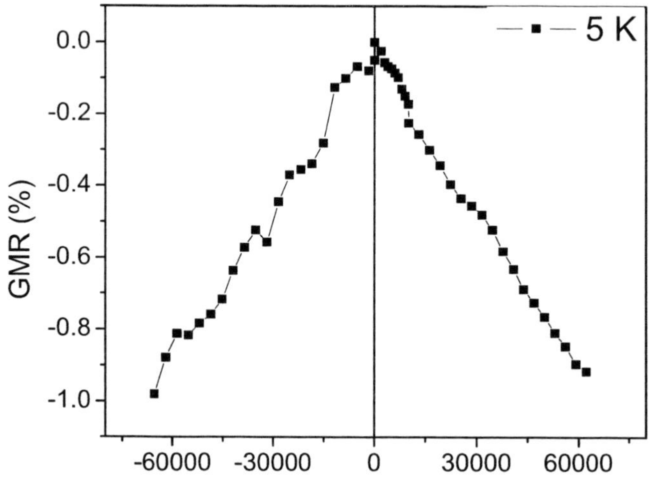

Figure 81. MR in as-prepared Cu_{50}-$(Fe_{69}Si_{10}B_{15}C_5)_{50}$ microwires measured at 5K.

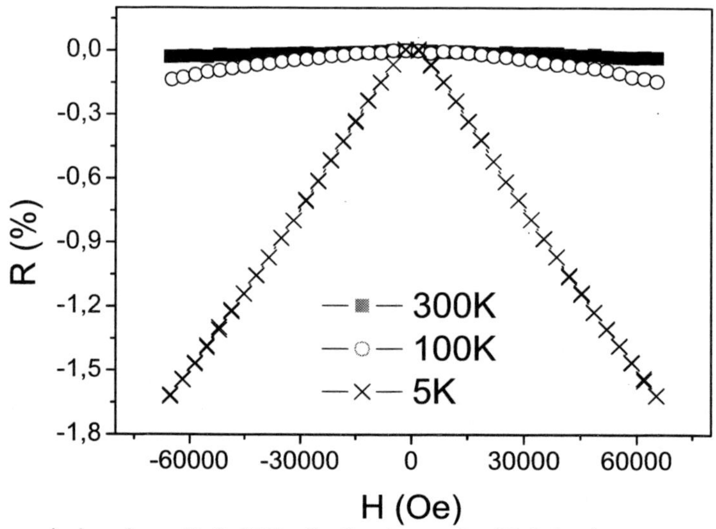

Reprinted with permission from Ref. [74], C. Garcia et al., "Fabrication, structural and magnetic characterization of thin microwires with novel composition $Cu_{70}(Co_{70}Fe_5Si_{10}B_{15})_{30}$" Journal of Alloys and Compounds 2008 (in press). Copyright Elsevier V. L. (2008) (figure 5).

Figure 82. Magnetoresistance of the as-cast microwires taken at 300, 100 and 5 K from -7 to 7 T. Magnetoresistance reaches significant values around 1.6 % in magnetic field H=7 T at 5 K.

4.4.3. Microwires with Mixed Structure

Glass-coated microwires with mixed structure (exhibiting 2 and/or more magnetic phases crystalline and/or amorphous phases) have been recently fabricated directly by the Taylor-Ulitovski method and/or by recrystallization of amorphous precursors [74,75]. Few peculiar properties, like hysteretic magnetic properties and magneto-resistance, MR, have been observed and studied. Particularly Cu_{50}-$(Fe_{69}Si_{10}B_{15}C_5)_{50}$ thin wire of few μm in diameter covered with a thin layer of Pyrex glass have been prepared by the Taylor-Ulitovsky technique [74].

Magnetization curves measured at different temperatures are shown in Figure 79 (a,b). In as-prepared sample the coercivity is of order of 5 Oe, although the hysteresis loop has the irregular shape typical for two-phase structure (see Figure 79b).

Magnetization measured under applied magnetic field (FC) and without magnetic field (ZFC) exhibit significant difference.

After heat treatments an increase of Hc is observed and the hysteresis loop shape change being more similar to the single phase alloy (Figure 80).

Like in the case of as-prepared microwires the magnetization measured under applied magnetic field (FC) and without magnetic field (ZFC) exhibit significant difference.

Analysis of the X-ray diffraction allows us to identify four different contributions: 1) Pyrex contribution at with broad maximum at $2\Theta \approx 22^o$; 2) Cu exhibiting 3 peaks at interatomic spaces, d, corresponding to 2.064, 1.792 and 1.273 A; 3) small amount of α-Fe with d=1.979Å and 4) amorphous halo.

Such crystalline structure allowed explaining the shape of the hysteresis loops observed in as-prepared state. Thus, magnetically soft amorphous phase and crystalline α-Fe phase result in irregular hysteresis loop typical for two-phase behaviour. The inhomogeneity of as-prepared samples consisting of crystalline paramagnetic Cu matrix with 2 ferromagnetic phases (Fe-rich amorphous and crystalline α-Fe) gives rise also to the significant difference in ZFC-FC curves. Additionally, such multi-phase structure led to appearance of magneto-resistance, MR in this sample. MR, of about 1% has been observed at 5K in studied s-prepared microwire (see Figure 81).

Similarly, magnetic and magneto-transport properties of $Cu_{70}(Co_{70}Fe_5Si_{10}B_{15})_{30}$ microwire have been studied. Coexistance of amorphous halo, *fcc* Cu and *hcp* Co phases has been detected by X-ray diffraction.

The *M(H)* curve (hysteresis loop) of the glass coating microwires at room temperature resembles superparamagnetic behaviour, previously observed in Cu-Co systems [75]. Similar features were observed at different measuring temperatures. The monotonic decrease observed above 100 K on *H(T)* and $M_r(T)$ generally fit previously predicted and experimentally observed dependences explained assuming coexistence of blocked and unblocked particles [75]. In real few factors can be taken into account in order to explain $H_c(T)$ dependence: 1) coexistence of blocked superparamgnetic grains related with complex crystalline structure; 2) thermal fluctuations; 3) internal stress arising from dependence of thermal expansions coefficients of the glass and metal.

Magnetization measured under 50 and 200 Oe applied magnetic field (FC) field and without magnetic field (ZFC) shows a similar behaviour. The blocking temperature was found around 25 K which corresponds to a critical diameter around 3.6 nm, which correspond with the value obtained by XRD.

Figure 82 shows the magnetoresistance response of the as-cast microwire measured at 5, 100 and 300 K. Relatively small and negative magnetoresistance has been observed at all temperatures. Decreasing the temperature MR increases from 0.05% till 1.6%.

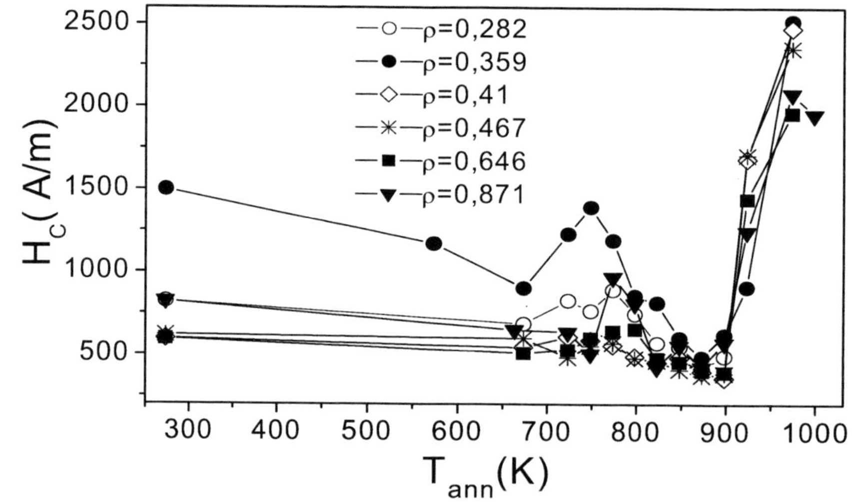

Reprinted with permission from Ref. [211], C. Garcia et al., "Soft Magnetic Behaviour of Nanocrystalline Fe-Based Glass-Coated Microwires", Journal of Optoelectronics and Advanced Materials, V8, No 5, (2006) pp.1667-1671. Copyright National Institute of Optoelectronics (2006) (figure 4).

Figure 83. Dependence of coercivity on the annealing temperature, T_{ann}, for $Fe_{71.8}Cu_1Nb_{3.1}Si_{15}B_{9.1}$ with the ratio *d*/*D* ranging from 0.282 to 0.646.

Summarizing, fabrication and characterization of a novel glass coated Cu_{50}-$(Fe_{69}Si_{10}B_{15}C_5)_{50}$ and $Cu_{70}(Co_{70}Fe_5Si_{10}B_{15})_{30}$ microwire with mixed structure prepared by the modified Taylor-Ulitovsky method have been reported. First structure, magnetic and magneto-transport characterization have been performed. Multi-phase structure consisting of Cu, small amount of α-Fe and amorphous phase has been observed. Magnetic and transport properties of Cu_{50}-$(Fe_{69}Si_{10}B_{15}C_5)_{50}$microwires depend on the annealing. MR effect is also observed.

4.5. Effect of Partial Crystallization and Nanocrystallization

Soft magnetic nanocrystalline alloys have attracted a great interest because such materials exhibit outstanding soft magnetic properties [31-38]. The nanocrystalline state has been achieved after partial devitrification of Fe-Cu-Nb-Si-B amorphous precursors using careful annealing of the amorphous precursors. After the first step of crystallization, FeSi or Fe crystallites are respectively finely dispersed in the residual amorphous matrix. In a wide range of crystallized volume fraction, the exchange correlation length of the matrix is larger than the average intergranular distance, d, and the exchange correlation length of the grains is larger than the grain size, D. Enhanced magnetic softness of Fe-rich nanocrystalline alloys was also attributed to a second complementary reason: the opposite sign of the magnetostriction constant of crystallites and residual amorphous matrix, which allows achieving of vanishing average magnetostriction constant.

Most attention has been paid to studies of nanocrystalline ribbons or wires, but recently great attention has been paid also to studies of nanocrystalline microwires [209-213].

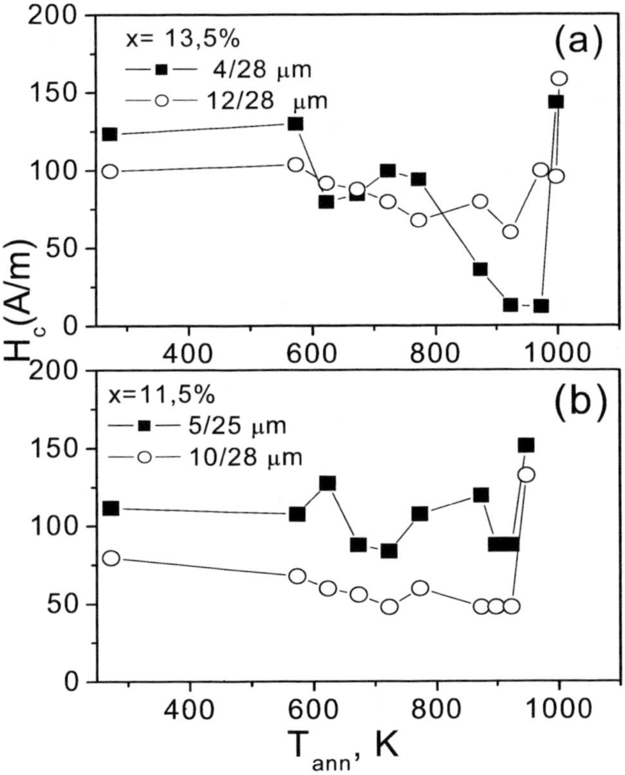

Figure 84. Dependence of coercivity on the annealing temperature, T_{ann}, for $Fe_{73.5}Cu_1Nb_3Si_xB_{22.5-x}$ microwires.

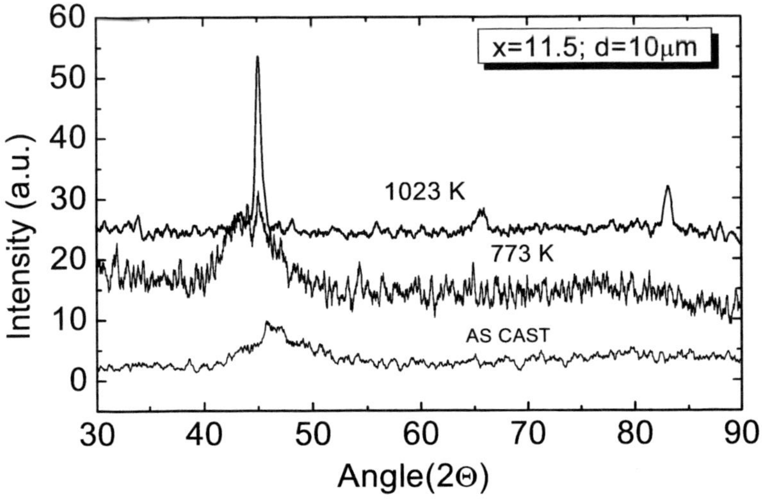

Figure 85. X-ray diffraction patterns of $Fe_{73.5}Cu_1Nb_3Si_xB_{22.5-x}$ microwires subjected to annealing at different temperatures.

4.5.1. Nanocrystalline Fe-Cu-Nb-Si-B Glass-coated Microwires

4.5.1.1 Magnetically Soft Nanocrystalline Microwires

Main attention has been paid to the studies of "*Finemet*"-type compositions microwires and effect of annealing on their soft magnetic properties.

As expected, magnetic properties of as-prepared and annealed $Fe_{73.4}Cu_1Nb_{3.1}Si_xB_{22.5-x}$ (x = 11.5, 13.5 and 16.5) and $Fe_{73.4-x}Cu_1Nb_{3.1}Si_{13.4+x}B_{9.1}$ ($0 \leq x \leq 1.1$) "*Finemet*"-type microwires significantly affected by their geometry [18, 209-211]. Figs 83 and 84 represent dependences of coercivity, H_c, on annealing temperature, T_{ann} for both "*Finemet*"-type microwires, $Fe_{73.4}Cu_1Nb_{3.1}Si_xB_{22.5-x}$ (x = 11.5, 13.5 and 16.5) and $Fe_{73.4-x}Cu_1Nb_{3.1}Si_{13.4+x}B_{9.1}$ ($0 \leq x \leq 1.1$).

Like in the case of Fe- based amorphous microwires, the coercivity of the sample $Fe_{72.3}Cu_1Nb_{3.1}Si_{14.5}B_{9.1}$ strongly increases as the ratio ρ decreases.

For both compositions of "*Finemet*"-type microwires the decrease of H_c has been, generally, observed at T_{ann} below 673 K, which could be ascribed to the internal relaxation stresses effect. In the range of annealing temperature of 673-723 K, a weak local minimum of H_c has been observed, with the temperature of that minimum depending on both alloy composition and geometry. Such a decrease of H_c could be ascribed to the structural relaxation of the material remaining the amorphous character such as has been widely reported in metallic glass alloys. A small relative hardening (increase of coercivity) can be observed after annealing around 723–773 K, which could be ascribed to the very beginning of the first stage of devitrification [2, 18, 31-38, 209-211]. It is interesting to note that the sample's geometry affects the value and the position of the local extremes on the $H_c(T_{ann})$ dependence (see Figs. 83,84). A deeper softening (optimum softness) with a rather low value of H_c is obtained in samples treated at T_{ann} = 773–873 K. The position of this minimum is also strongly affected by the sample geometry (see Figs. 83,84). This magnetic softening is related to the nanocrystallization process owing the precipitation of fine grains (10–15 nm) of α-Fe(Si) phase within the amorphous matrix. Such interpretation has been confirmed by the X-ray diffraction studies of as-prepared and annealed at different temperatures samples [209,210] (see Figure 85). As can be deduced from Figure 85, amorphous state was observed for the as-prepared samples $Fe_{73.4}Cu_1Nb_{3.1}Si_xB_{22..5-x}$ with x =11.5. The X-ray diffraction patterns for the annealed samples with x = 11.5 indicate the appearance of a-Fe(Si) crystalline grains (randomly textured) for annealing temperature, T_{ann} above 773 K (see Figure 85). Annealing above 973 K causes the appearance of new boride phases. The mean grain diameter of precipitated crystallites are estimated using the Debye-Sherrer equation is 2; 12 and 14 nm for T_{ann}= 773, 823 and 1023 K respectively [210].

Excellent magnetic softness (coercivity of about 15 A/m) has been realized in $Fe_{73.5}Cu_1Nb_3Si_xB_{22.5-x}$ microwires with x = 13.5 [209,210] at appropriate annealing conditions (see Figure 84).

Optimum softest behaviour, that is, a deeper softening with very low value of coercivity is obtained in the samples treated 773–873 K which could be ascribed to the fact that the first crystallization process has been developed, leading to fine nanocrystals α-Fe (Si) of grain size around 10 nm with 70–80% of relative volume, such it has been widely reported for *Finemet* ribbons (see refs. [31-38]). An abrupt increase of the coercivity is shown by the samples treated at temperature above 873 K, indicating the beginning of such increase should be connected to the precipitation of iron borides (with grain size larger than 50 nm) which implies a magnetic hardening character. The beginning of the increase of H_c varies depending mainly on the sample composition as well as on the geometry.

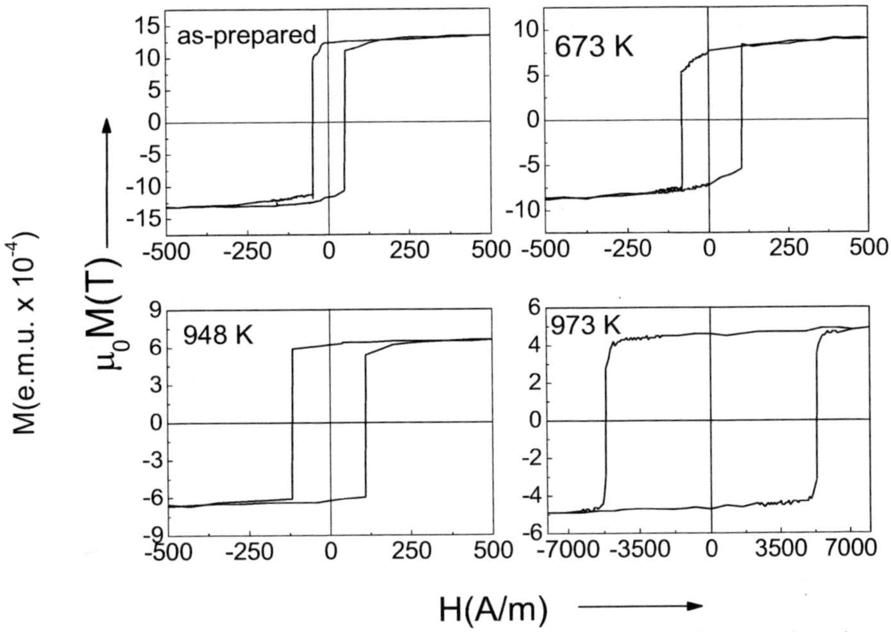

Reprinted with permission from Ref. [136] A. Zhukov, "Glass-coated magnetic microwires for technical applications" J. Magn and Magn, Mater. 242-245, 216 (2002) Copyright Elsevier Science B.V. 2002 (figure 4b).

Figure 86. Hysteresis loops of $Fe_{71.8}Cu_1Nb_{3.1}Si_{15}B_{9.1}$ in as-prepared and annealed at different temperature microwires.

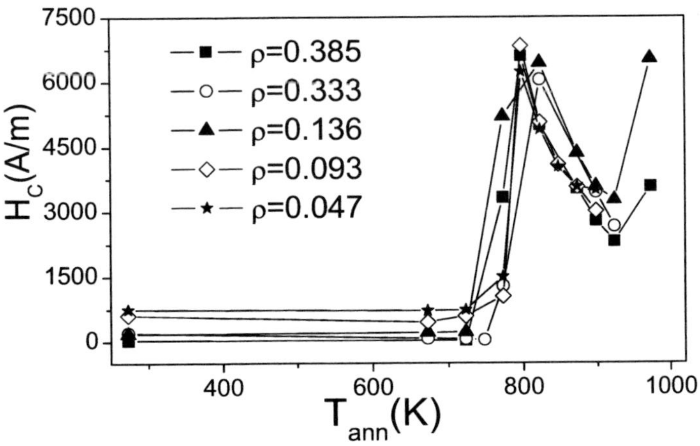

Reprinted with permission from Ref. [210] J. Gonzalez et al., "High Coercivity of Partially Devitrified Glass-Coated *Finemet* Microwires: Effect of Geometry and Thermal Treatment", IEEE Trans. Magn., V 36, Nº 5, 2000. Copyright IEEE Magnetic Society (2000) (figure 2).

Figure 87. Coercivity dependence with the annealing temperature of $Fe_{72.3}Cu_1Nb_{3.1}Si_{14.5}B_{9.1}$ microwire.

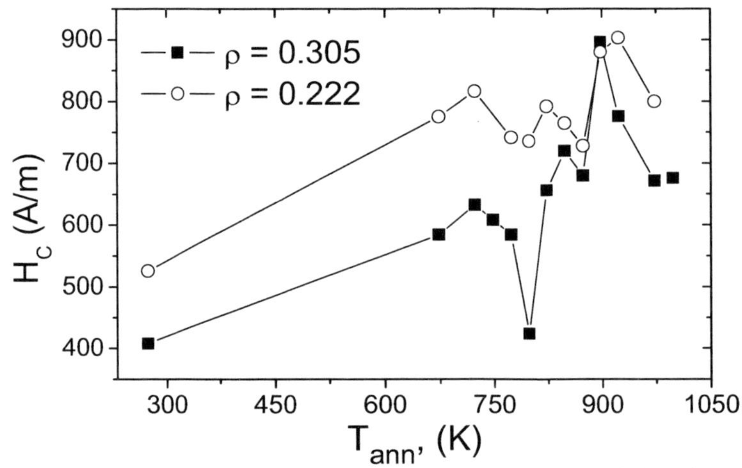

Reprinted with permission from Ref. [210] J. Gonzalez et al., "High Coercivity of Partially Devitrified Glass-Coated *Finemet* Microwires: Effect of Geometry and Thermal Treatment", IEEE Trans. Magn., V 36, Nº 5, 2000. Copyright IEEE Magnetic Society (2000) (figure 3).

Figure 88. Dependence of the coercivity of $Fe_{73.4}Cu_1Nb_{3.1}Si_{13.4}B_{9.1}$ microwire on the annealing temperature.

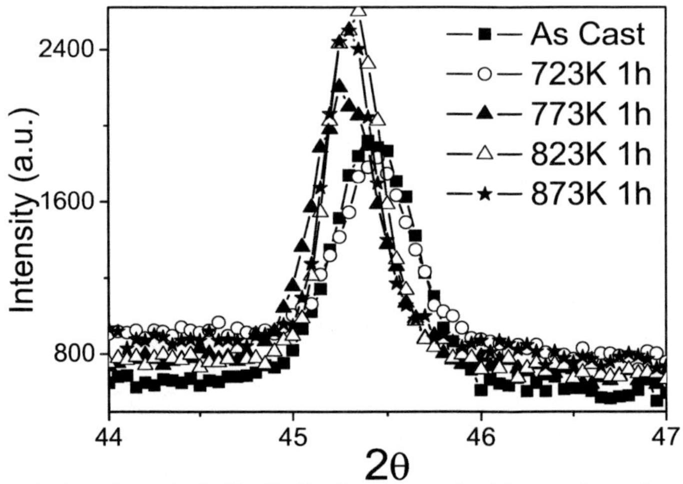

Reprinted with permission from Ref. [212] C. García et al., "Stress dependence of coercivity in nanocrystalline Fe79Hf7B12Si2 glass-coated microwires", J.Appl. Phys. 99 (2006) 08F116. Copyright (2006), with permission from AIP (figure 1).

Figure 89. Evolution of the main peak corresponding to (110) reflection of the XRD patterns of the $Fe_{79}Hf_7B_{12}Si_2$ glass-coated microwire with annealing.

Observed above effect of the thickness of glass coating has been interpreted as the consequence of both strong internal stresses and the difference in the thermal exchange conditions during fabrication and annealing due to isolating properties of the glass coating. As a consequence, the coercivity of the as-prepared and treated (573–973 K) samples results to be very sensitive to the *d/D* ratio with significant difference in the softening (nanocrystallization) and hardening (second crystallization process) character.

4.5.1.2. Semihard Magnetic Nanocrystalline Microwires

In fact, annealed at elevated temperatures (above 873 K) microwires present an abrupt increase of H_c, indicating the beginning of the precipitation of second crystalline phase–iron borides (with grain size larger than 50 nm). Such beginning of the increase of H_c varies depending mainly on sample composition as well as geometry (Figs 83,84). It must be noted that for the thickest glass coating the increase of H_c appears at lower temperature, which could be related to the fact that the internal stresses induce some ordering to hinder the crystallization. It is important that the hysteresis loop in certain microwire compositions remains roughly rectangular in the whole range of annealing temperatures for certain compositions (Figure 86) [18]. Such temperature dependence of the hysteresis loops allows one to tailor the coercivity of microwires with rectangular hysteresis loops in a wide range of coercivities. This is in fact quite important for some particular applications of glass-coated microwires, related with magnetic encoding using magnetic TAGs. The effect of annealing temperature on the coercivity, H_c, of the alloy $Fe_{72.3}Cu_1Nb_{3.1}Si_{14.5}B_{9.1}$ with ratio ρ as parameter is shown in Figure 87. A sharp magnetic hardening (increase of coercivity) can be observed after annealing around 773-823 K, i.e. much before the crystallization (second phase precipitation).

It should be indicated that $H_c(T_{ann})$ dependence is strongly affected by the microwires composition. Thus in slightly different composition $Fe_{72.3}Cu_1Nb_{3.1}Si_{14.5}B_{9.1}$ first maximum on $H_c(T_{ann})$ dependence takes place at the same range of the annealing temperature as in $Fe_{73.4}Cu_1Nb_{3.1}Si_xB_{22.5-x}$ (x = 11.5 and 13.5), but the magnetic hardening is much stronger (compare Figure 87 with Figs. 83,84) [18, 209-211]. This strong magnetic hardening of $Fe_{72.3}Cu_1Nb_{3.1}Si_{14.5}B_{9.1}$ microwire is followed by new magnetic softening with an increase of T_{ann} above 823 K and consequent new magnetic hardening at T_{ann} > 923 K which is accompanied by deterioration of mechanical properties (Figure 87).

It should be noted that quite different $H_c(T_{ann})$ behaviour has been observed for very similar composition ($Fe_{71.8}Cu_1Nb_{3.1}Si_{15}B_{9.1}$) and $Fe_{73.4}Cu_1Nb_{3.1}Si_xB_{22.5-x}$ (x = 11.5 and 13.5), when just local maximum has been observed at about 750 K (see Figs.82,83).

In the case of $Fe_{73.4}Cu_1Nb_{3.1}Si_{13.4}B_{9.1}$ samples such strong magnetic hardening is not observed (Figure 88) although $H_c(T_{ann})$ is also different from $Fe_{73.4}Cu_1Nb_{3.1}Si_xB_{22.5-x}$ (x = 11.5 and 13.5). In order to understand such considerable difference in $H_c(T_{ann})$ in very similar compositions X-ray analysis of $Fe_{72.3}Cu_1Nb_{3.1}Si_{14.5}B_{9.1}$ has been performed. X-ray analysis of $Fe_{72.3}Cu_1Nb_{3.1}Si_{14.5}B_{9.1}$ microwires does not detect a presence of crystalline phases in as-prepared state as well as after annealing at 823 K. On the other hand, transmission electron microscopy (TEM) diagrams of the sample annealed at 823K allows one to detect a small amount of fine grains of α-Fe, γ-Fe, and α-Fe(Si) [18]. Observed structural features permit one to correlate observed magnetic hardening with precipitation of fine crystallites. Such a difference in magnetic behavior with conventional *Finemet*-type microwires can be attributed to the slightly different alloy compositions, different size of precipitating nano-grains as well as to the effect of high internal stresses. It is well known that the best magnetic softness is achieved when the nanocrystalline structure consisting of small α-Fe(Si) grains and amorphous matrix has a vanishing magnetostriction constant. It is possible that even a small change of the alloy composition does not permit one to achieve such a vanishing magnetostriction constant. On the other hand, strong internals stresses due to glass coating can result in a change of structure of the precipitating fine grains. Thus, strong internal stresses (about 100 MPa or even more) are mainly induced by the difference in the thermal expansion coefficients of the glass and the metallic nucleus [2-4]. It is well known that internal strains of different nature can be the origin of martensite-type transformation in alloys of Fe. Probably, the strong internal stresses induce a precipitation of the γ-Fe fine grains instead or apart from α-Fe(Si) grain during the first stage of the crystallization process. Particularly, the presence of crystalline γ-Fe could be the reason for the observed magnetic hardening [210]. As a consequence, the origin of such strong magnetic hardening at low annealing temperatures without deterioration of mechanical properties could be ascribed to some peculiarities of the first recrystallization process under the effect of strong internal stresses induced by glass coating and differences in the alloy composition.

Generally all samples of $Fe_{72.3}Cu_1Nb_{3.1}Si_{14.5}B_{9.1}$ microwires have similar $H_c(T_{ann})$ dependencies, exhibiting 2 minimums: one at around 650 K, the other one – between 800 and 900 K. The annealing temperatures, T_{ann}, corresponding to the magnetic hardening and softening as well as the value of the coercivity at both temperatures depend on the ρ–ratio for each composition. Usually, largest coercivity values and lowest annealing temperatures of the maximum and the minimum correspond to the lowest ρ values. $Fe_{73.4}Cu_1Nb_{3.1}Si_{13.4}B_{9.1}$ microwires do not exhibit strong magnetic hardening (see Figs 83,84).

In any case further studies are needed I order to clarify such different $H_c(T_{ann})$ in slightly different *Finemet* –type microwires compositions.

4.5.2. Nanocrystalline Structure in FeHfBSi Glass-coated Microwire

Since Zr has some chemical interaction with the glass the other composition of nanocrystalline glass-coated microwires fabricated by the modified Taylor-Ulitovski method, $Fe_{79}Hf_7B_{12}Si_2$, has been introduced. The total external diameter of the studied microwire was 24 μm and the one of the metallic nucleus was 15 μm. These samples have been annealed in the temperature range 723-873 K during 1 hour [212-214].

The evolution of the first peak of XRD with the annealing temperature is presented in Figure 89. As it can be seen, the position and width of such peaks change significantly with the annealing. It is worth to mention that even the as-cast sample exhibits a slightly crystalline structure. The average grain size estimated from the peak width by means of Debye-Scherrer equation is about 17 nm [212, 213].

The evolution of the average grain size of nanocrystals with the annealing temperature is presented in the Figure 90. The grain size of the nanocrystals increases from about 17 nm up to 35 nm after annealing at 873 K.

Rather elevated coercivity, H_{c0}, has been observed in these microwires. The changes of the coercive field, H_{c0}, with the annealing temperature are shown in the figure 91. As can be seen, a significant enhancement of the soft magnetic character takes place in the range 773-873 K, which results to coincide with the increase of the grain size, D (see Figure 91). Nevertheless, the values of $H_{c0} \approx 600$ A/m and $D \approx 30$ nm for these annealed samples (773-873 K) are quite larger than those reported in the classical *Finemet* Fe-rich nanocrystalline ribbon shaped ($H_c \approx 1$ A/m and $D \approx 10$ nm). These differences could be associated to the strength and complexity of the internal stresses actuating on the metallic nucleus due to the glass coating. Such internal stresses arise some ordering to hinder the nanocrystallization process as it has been previously observed in glass-coated *Finemet*-type microwires [17].

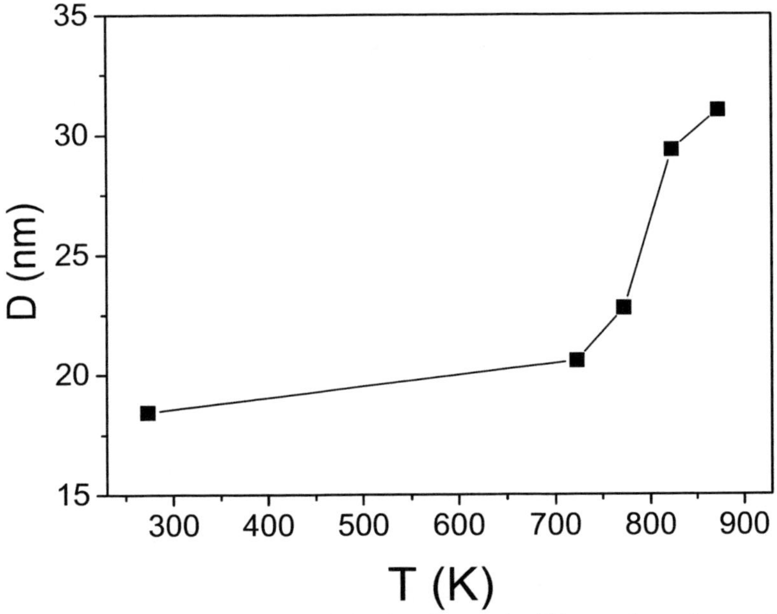

Reprinted with permission from Ref. [212] C. García et al., "Stress dependence of coercivity in nanocrystalline Fe79Hf7B12Si2 glass-coated microwires", J.Appl. Phys. 99 (2006) 08F116. Copyright (2006), with permission from AIP (figure 2).

Figure 90. Evolution of the average grain size of $Fe_{79}Hf_7B_{12}Si_2$ glass coated microwire under annealing.

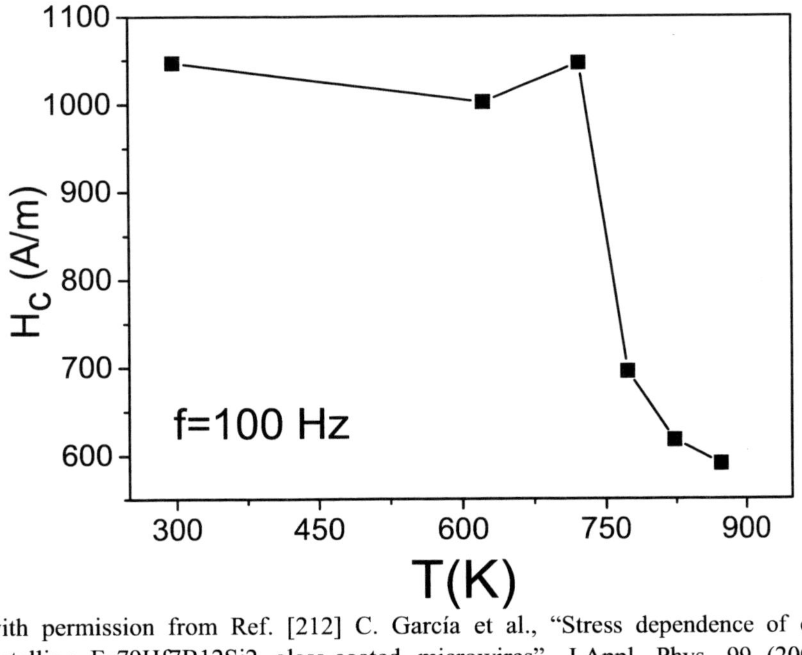

Reprinted with permission from Ref. [212] C. García et al., "Stress dependence of coercivity in nanocrystalline Fe79Hf7B12Si2 glass-coated microwires", J.Appl. Phys. 99 (2006) 08F116. Copyright (2006), with permission from AIP (figure 4).

Figure 91. Evolution of the coercivity of $Fe_{79}Hf_7B_{12}Si_2$ glass coated microwire under annealing.

Figure 92 presents the stress dependence of $\frac{H_{c0}-H_{cS}}{H_{cS}}$ (H_{c0} is the coercivity without stress and H_{cS} is the coercivity under stress) with the applied stress, σ_a, for as-cast and annealed samples. It can be seen that in the as-cast and annealed samples at low annealing temperatures (< 773 K) a decrease of H_c with σ_a, while in samples treated above 773 K, H_c increases with σ_a. Such behaviour could result from the different magnetostrictive character of the samples. In fact, the effective saturation magnetostriction in soft nanocrystalline alloys has been assumed to be a balance of two main opposite contributions, namely: the first one arising from the partial volume of nanocrystalline phases and the second one arising from the residual amorphous matrix [18]. Because of the content and the distribution of the nanocrystalline and amorphous phases change with the annealing temperature, different stress dependence of the coercivity can be expected. In the as-cast and low temperatures annealed samples (T_{ann} < 773 K) the coercivity monotonically decreases with the applied tensile stress, which should be connected to the positive magnetostrictive character of these samples, while in the samples annealed above 773 K the coercivity behaves in an opposite way, being increased with σ_a, as a consequence of the negative effective magnetostriction in the annealed samples, characterized by the massive presence of nanograins.

The grain size plays a relevant role to explain the coercivity behaviour. In fact, an increase of the grain size (15-35 nm) with the simultaneous decrease of the coercivity is observed. In addition, this dependence of coercivity with the grain size seems to reveal some kind of deviation from the random anisotropy model proposed by Herzer for the nanocrystalline Finemet material [31-38]. According to such model, an enhanced magnetic softness correlates with the grain size and the best magnetic softness is achieved under conditions of averaging grain size, typically of 10-15 nm, is much below than the exchange correlation length, being around 35-40 nm.

These differences could be newly ascribed to the strength and complexity of the internal stresses acting on the metallic nucleus due to the glass coating. Another reason to explain the relationship between the grain size and the coercivity obtained for these thermally treated microwires could be assigned to the difference of the volume fraction of the α-Fe phase. In fact, the devitrification starts above 723 K (see Figure 91). Moreover, the crystalline volume fraction and the size of the crystallites increase above 723 K. Consequently, the exchange coupling occurs almost directly between crystals and the random anisotropy model efficiently works to decrease the coercivity. However, in the annealed samples below 723 K, the crystalline phase is embedded in the amorphous phase and the coupling occurs through the last one and, therefore, the observed behaviour of H_c is mainly due to the amorphous phase.

The problem of the surface/interface magnetic behaviour requires however further studies which in particular should include measurements at temperatures above Curie point of the amorphous matrix where only ferromagnetic grains should contribute to the effective magnetostriction simplifying a separation between bulk and surface

Some aspects of magnetic behavior of glass-coated magnetic microwires with nanocrystalline structure have been overviewed. Both magnetically soft and hard behavior is analyzed. Special attention has been paid to the geometry (total diameter and diameter of the metallic nucleus) and compositions (FeCuNbSiB and FeHfBSi). The role of the complex internal stresses arising from the different dilatation coefficient between the metallic nucleus and the insulating sheath results to be as one of the most important factor to understand the soft magnetic properties which present these nanocrystalline microwires. The magnetic behaviour is overviewed in terms of the nanosize dimensions and evolution of the nano-grain size with annealing temperature.

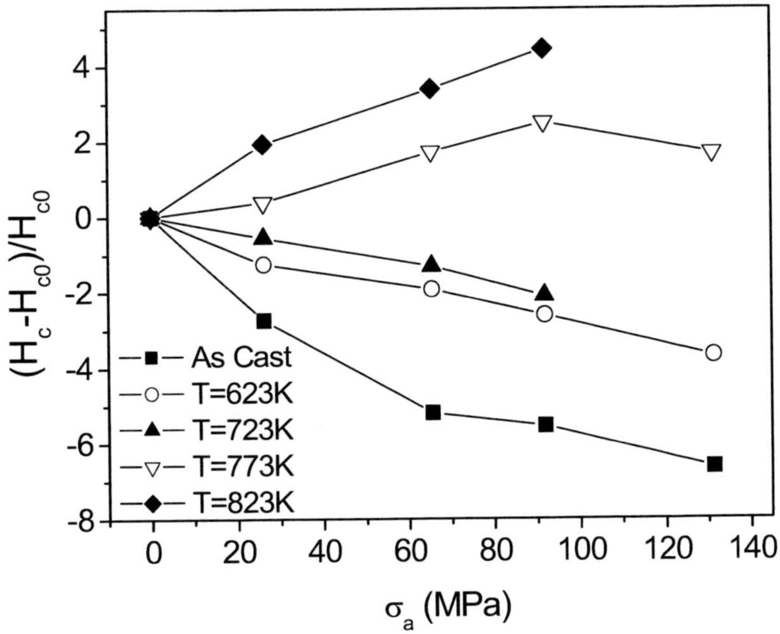

Reprinted with permission from Ref. [212] C. García et al., "Stress dependence of coercivity in nanocrystalline Fe79Hf7B12Si2 glass-coated microwires", J.Appl. Phys. 99 (2006) 08F116. Copyright (2006), with permission from AIP (figure 5).

Figure 92. Stress dependence of the coercivity of $Fe_{79}Hf_7B_{12}Si_2$ glass coated microwire in as-prepared state and after annealing at different temperatures.

It should be noted that recently high frequency propertics, such as GMI effect and frequency dependence of magnetic permeability have been studied in Finemet tipe glass coated microwires [214]. Magnetic properties have been measured and compared to amorphous Co-based microwires. Two alloys of the Finemet family ($Fe_{71.1}Nb_3Cu_{1.2}B_{7.9}Si_{16.8}$ and $Fe_{73.2}Nb_{2.9}Cu_{1.2}B_{6.9}Si_{15.8}$) of this kind of microwires have been studied. Dependence of the magnetic anisotropy with both the composition and the diameters of the microwires were observed. Magnetic permeability has been measured within the 200 MHz to- 4 GHz range up to 623 K. Fair stability of the microwave properties and their reversible behaviour has been observed.

Chapter 5

NON –MAGNETICALLY SOFT MICROWIRES

5.1. MICROWIRES WITH GRANULAR STRUCTURE. MAGNETO-RESISTANCE EFFECT

Granular systems formed by inmiscible elements (Co, Fe, Ni)-(Cu, Pt, Au, Ag) have been introduced few years ago [215-218]. Such immiscible alloys exhibit unusual magnetic properties such as the GMR effect related with mixed ferromagnetic-paramagnetic microstructure and scattering of the electrons on grain boundaries between these two phases. A rapid quenching technique such as melt spinning has been employed recently for fabrication of Co-Cu ribbons [218]. The high quenching rate typical for the melt spinning method has been used for fabrication of metastable crystalline materials when the crystalline phases, stable at high temperature, can be quenched-in at room temperature. Particularly, super-saturated solid solutions can be obtained at room temperature after rapid quenching, because the solubility of Co in cu changes with temperature (decreasing with decreasing of the temperature). These super-saturated solid solutions are metastable at room temperature and therefore thermal treatment results in decomposition of such structure and precipitations of the small Co grains in Cu matrix. In this case the GMR and enhanced coercivity were obtained generally after recrystallization of such metastable phases.

Consequently, the Taylor-Ulitovsky technique has been also employed to obtain granular glass-coated microwires (km long) [5, 58, 59, 71-74, 219] from inmiscible elements. Such granular structure has been achieved mainly by decomposition of metastable crystalline phases. High quenching rate is also useful for fabrication of metastable crystalline materials when the crystalline phases, stable at high temperature, can be quenched-in at room temperature.

The experimental data on X-ray diffraction of microwires from inmiscible elements are generally in agreement with those previously reported for ribbons with small Co concentrations, when single phase peaks have been observed at $x<10\%$. The presence of two phases has been deduced from careful analysis of the maxima for $x>10\%$, with interatomic spacings of 0.209 nm (Cu) and 0.206 nm (Co). Nevertheless, it is important to mention that the coexistence of 2 phases is found even for $x>35\%$.

Like in Co_x-Cu_{100-x} microwires with $x > 10\%$, the XRD analyses for the as-prepared Cu-Fe microwires showed two maxima in the diffraction peak with the lowest diffraction angle corresponding to the Cu and the bcc α-Fe phases.

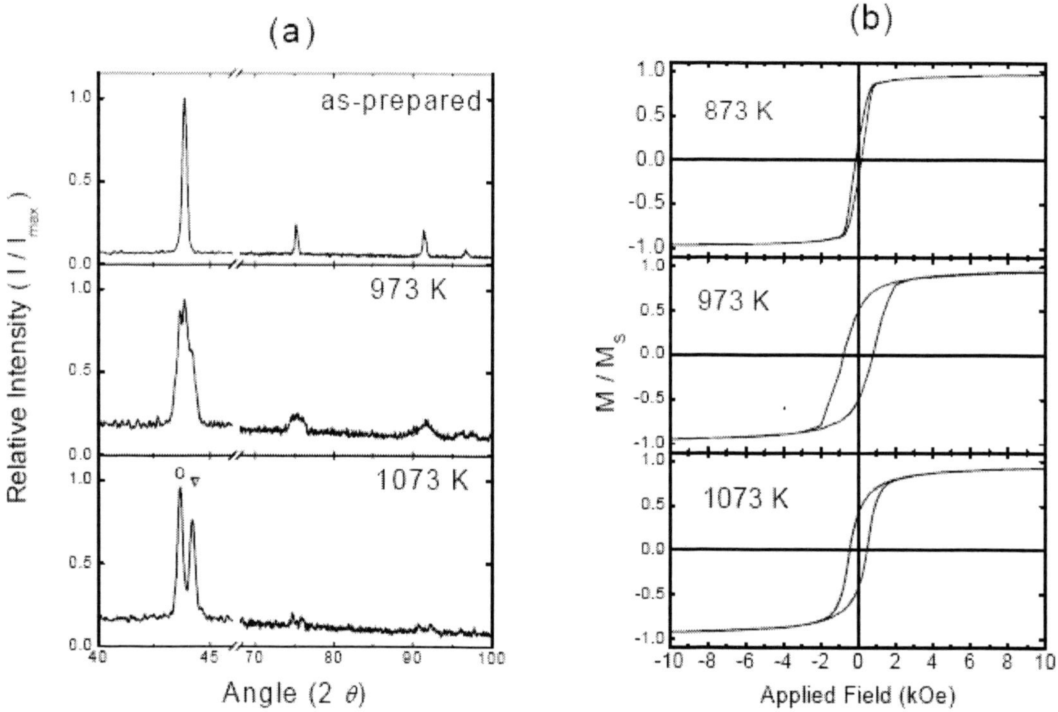

Reprinted with permission from Ref. [5] A. Zhukov et al "Magnetoresistance in thin wires with granular structure", J. Magn. and Magn. Mater. 294 (2005) 165-173. Copyright Elsevier (2005) (Figure 1).

Figure 93. Evolution of the X-ray patterns and hysteresis loops of $Co_{29}Ni_{25}Mn_1Cu_{45}$ microwires measured at room temperature after different annealing temperatures.

As mentioned above, in some cases this granular structure is metastable at room temperature and the consequent heat treatment results in the re-crystallization of this structure with the formation of stable phases. At certain conditions of such heat treatment this new granular structure can exhibit interesting magnetic and transport properties.

As an example, Figure 93 (a) shows the X-ray diffraction patterns for the $Co_{29}Ni_{25}Cu_{45}Mn_1$ sample, in its as-prepared state and after thermal treatment at two different temperatures T_{ann}. In the as-prepared microwire, quenched from a high temperature, only a single phase, with a lattice parameter *a* =0.3573 nm, was found (Figure 93a). When the sample was heated at high temperatures, this metastable phase started to decompose. This can be seen in the sample treated at 973 K, where the segregation mechanism is not yet complete (Figure 93a). After annealing at 1073 K the sample shows the equilibrium phases, consisting of a copper-rich matrix (*a* =0.3591 nm) with Co-rich particles (*a* = 0,3545 nm) (Figure 93a) [58].

In the Cu-Ni-Co system with more then 20% Ni, a continuous series of face centered cubic solid solution exist in high temperatures [58, 220, 221]. When the alloy is quenched from the melt, the metastable cubic phase is found in room temperature. Heat treating these samples, the segregation process takes place and the metastable phase starts to decompose into a paramagnetic Cu -rich phase with larger lattice parameter that forms the paramagnetic matrix, and a fine grained ferromagnetic Co rich phase with smaller lattice parameter. At this stage the coercivity rises quickly. Increasing the annealing temperature, the equilibrium of phases is reached more rapidly, the grains start to grow and the coercivity drops down. These results are in agreement with the magnetic hardening mechanism in bulk CuNiCo permanent magnet alloys, and the maximum value of coercivity obtained after the annealing at 973 K for one hour also agrees with the one reported in the literature [58].

Magnetization curves measured at different annealing temperatures are shown in Figure 93 b. In as-prepared sample the coercivity is of order of tens of Oe. After heat treatments a substantial increase of the coercivity, H_c, is observed at annealing temperatures, T_{ann}, above 773K (see Figure 93b). Increasing of H_c from 80 up to 750 Oe (at T_{ann}=973K) in $Co_{29}Ni_{25}Mn_1Cu_{45}$ has been found.

Such magnetic hardening should be attributed to the decomposition of the metastable Cu-rich Cu-Ni-Co solid solution and precipitation of Co or Ni crystallites from the metastable Cu-rich solid solution (see, for instance ref [2]). The single-phase solid solution is unstable at room temperature and the annealing results in its decomposition, as can be appreciated in Figure 93a.

Temperature dependence of the hysteresis loops of as-prepared and annealed at 973 K samples is also quite different. The coercivity, H_c, and remanent magnetization, M_r/M_s, exhibit a minimum in the as-prepared sample at about 120 K, while saturation magnetization, M_s, exhibits some increasing at low temperatures (see Figure 94).

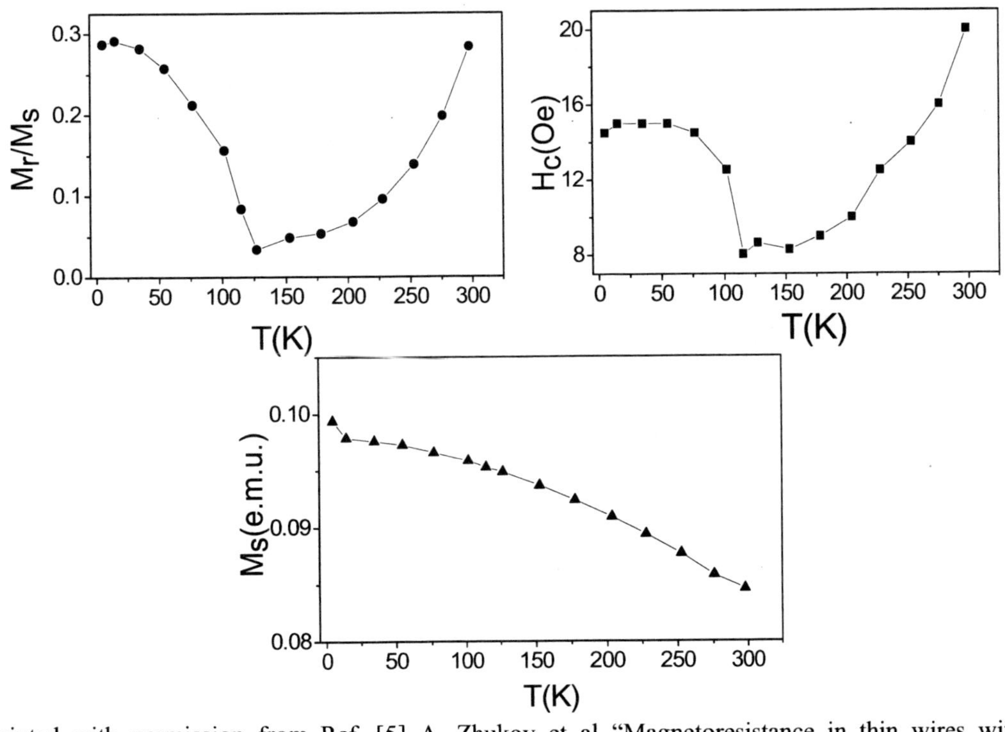

Reprinted with permission from Ref. [5] A. Zhukov et al "Magnetoresistance in thin wires with granular structure", J. Magn. and Magn., Mater. 294 (2005) 165-173. Copyright Elsevier (2005) (Figure 2).

Figure 94. Effect of temperature on hysteretic magnetic properties of as-prepared $Co_{29}Ni_{25}Mn_1Cu_{45}$ sample.

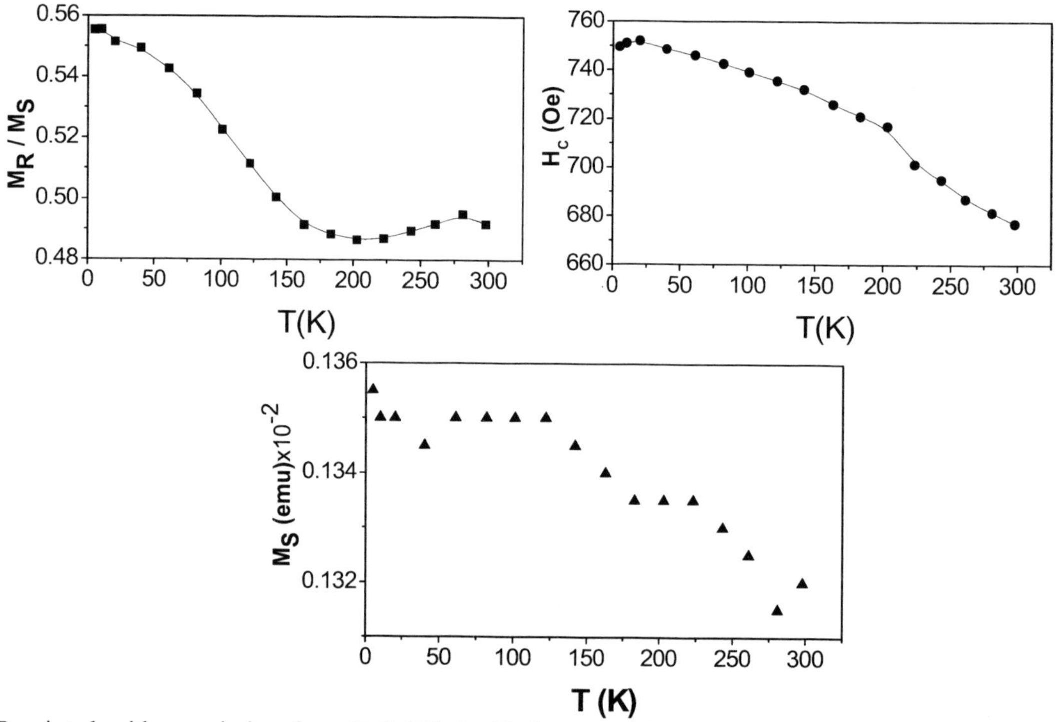

Reprinted with permission from Ref. [5] A. Zhukov et al "Magnetoresistance in thin wires with granular structure", J. Magn. and Magn., Mater. 294 (2005) 165-173. Copyright Elsevier (2005) (Figure 3).

Figure 95. Effect of temperature on hysteretic magnetic properties of annealed at 973 K $Co_{29}Ni_{25}Mn_1Cu_{45}$ sample.

In contrast, H_c of annealed sample increases decreasing the temperature at 10-300 K, showing slight maximum at around 20K, while M_r/M_s increases decreasing the temperature below 150 K (see Figure 95). Such increase of coercivity at low temperatures observed in as-prepared sample can be attributed to an increase of the ferromagnetic order contribution at low temperatures.

MR response has been measured in as-prepared and treated at 973 K samples. Considerable magneto-resistance, MR, has been found in both microwires (see Figs 96-98) in the temperature range of 5-300 K. MR effect of treated sample achieves 3% at 5K under magnetic field, H, of about 100 kOe. Increasing the temperature, MR effect of annealed sample decreases, especially above 120K. On the other hand, MR response of as-prepared sample becomes higher at room temperature (see overview of the temperature dependence of resistance under or without magnetic field in Figure 97).

Magnetic field dependence of MR in as-prepared and annealed samples is quite different, especially in the range of low fields (see Figure 96-98). Roughly monotonic decay is observed in as-prepared microwires at H>1000 Oe (see Figure 96). Annealed sample exhibits a maximum and considerable hysteresis on magnetic field dependence of MR at low magnetic fields. The position of this maximum correlates well with the values of coercivity, H_c, obtained from magnetization curves. Such peculiar shape of the magnetic field dependence of MR with positive longitudinal MR at low fields and negative at higher magnetic fields can be attributed to the contribution of both anisotropic and giant magnetoresistance. In this case the anisotropic contribution should be attributed to the ferromagnetic nature of Co or Ni-rich phases while the negative longitudinal MR should be related with the single domain ferromagnetic particles embedded in an immiscible metallic medium [58].

It is worth mentioning that the temperature at which coercivity has a minimum in as-prepared sample (about 120 K) coincides to the temperature at which the MR of as-prepared sample begins to decrease (see Figure 97). In order to explain such behavior, neutron diffraction data have been analyzed in the temperature range between 4 and 300 K. Although before it was suggested that room temperature X-ray peak corresponds to the single cubic phase with lattice parameter a=3.573 Å, more detailed analysis of the neutron diffraction spectra allow to assume that the neutron diffraction peaks should be considered as double, formed by the peak (111) of Cu FCC (lattice parameter 3.615Å) and by the peak (111) of Co FCC (lattice parameter 3.544Å) (Figure 99 a,b).

Another system of microwires made from inmisible alloys is Co-Cu. From X-ray diffraction of $Co_{10}Cu_{90}$ microwires the following phases have been observed [2]:

a) three peaks of the FCC Cu phase (lattice parameter: 3.61 Å) corresponding to (111), (011) and (010) reflections (linked to atomic spacings of 2.09, 1.81 and 1.28 Å)
b) three peaks of the HCP Co phase (lattice parameters: 2.51 and 4.07 Å) corresponding to (002), (011) and (110) reflections (linked to atomic spacings of 2.05, 1.77 and 1.25 Å).

The sample's geometry affects strongly the structure. Thus the intensity of the X-ray lines that correspond to the Co grains decrease with increasing the ratio ρ. It is well-known that the strength of the internal stresses increases increasing the glass coating thickness [2]. In this way the crystallization process in the presence of the internal stresses probably develops easier. It means that the internal stresses act as the reinforced factor and changing the glass coating thickness one can tailor the crystalline structure of the glass coated microwires. It should be noted that previously it was found the presence of even only single phase with a lattice parameter a = 0.36 nm in $Co_{10}Cu_{90}$ microwire with fixed geometry, while in the other compositions, with x>10%, two phases were detected [2]. Similarly it was found by the XRD method that, in the as-quenched Co_xCu_{100-x} ribbons, single phase *fcc* peaks appear for x<20% [2, 58, 59] and x>35%. Since the lattice parameter was larger than for pure Co (a=0.35447 nm) and smaller than that of pure Cu (a=0.3615 nm), same dissolution of Co atoms in the Cu matrix has been proposed.

The hysteresis loop measured at room temperature is presented in Figure 100 a. As can be observed, the saturation does not achieved even at highest magnetic field. The coercivity, H_c, is low enough (about 40 Oe). The sample annealing in the conventional furnace gives rise to the increase of H_c up to 200 Oe (Figure 100 b). Such phenomenology should be explained considering change of the size and morphology of Co precipitates after annealing (these studies are now under the progress). It is reasonable to suppose that the annealing results in the precipitation of a new Co grains from the Cu-rich solid solution due to low solubility of Co in Cu matrix at room temperature and growth of Co precipitates [2, 58,59].

The magnetization measured under applied magnetic field (FC) and without magnetic field (ZFC) exhibit significant difference for as-prepared sample (see Figure 101). Such difference confirms the presence of small ferromagnetic grains in paramagnetic matrix.

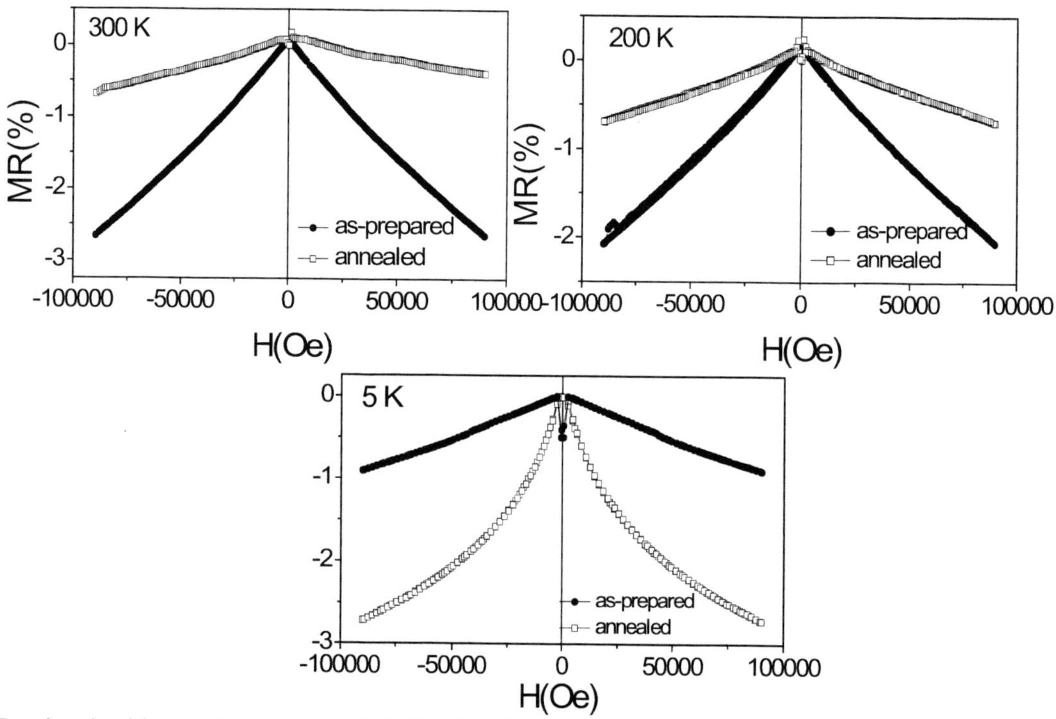

Reprinted with permission from Ref. [5] A. Zhukov et al "Magnetoresistance in thin wires with granular structure", J. Magn. and Magn., Mater. 294 (2005) 165-173. Copyright Elsevier (2005) (Figure 4).

Figure 96. Magnetic field dependence of MR measured in as-prepared and annealed at 973 K microwires measured at 5, 200 and 300K.

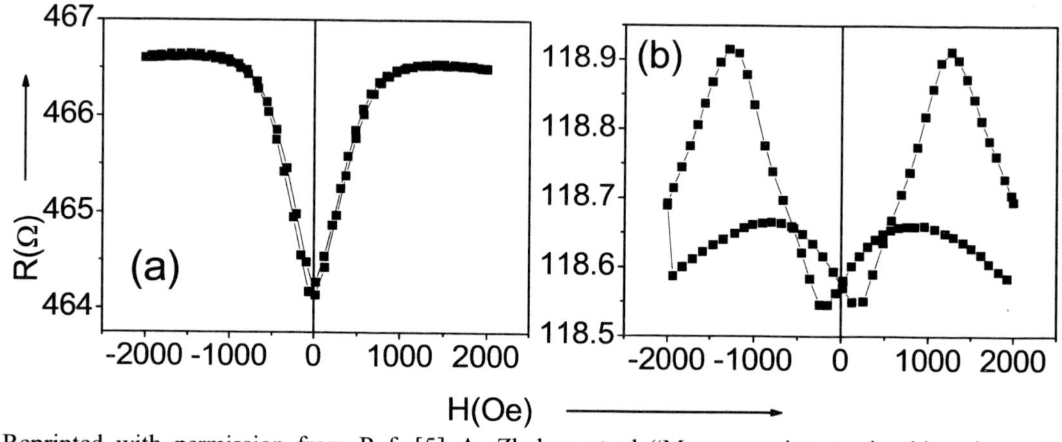

Reprinted with permission from Ref. [5] A. Zhukov et al "Magnetoresistance in thin wires with granular structure", J. Magn. and Magn., Mater. 294 (2005) 165-173. Copyright Elsevier (2005) (Figure 5).

Figure 97. Magnetic field dependence of MR measured in as-prepared (a) and annealed at 973 K (b) microwires measured at 5K at low magnetic field.

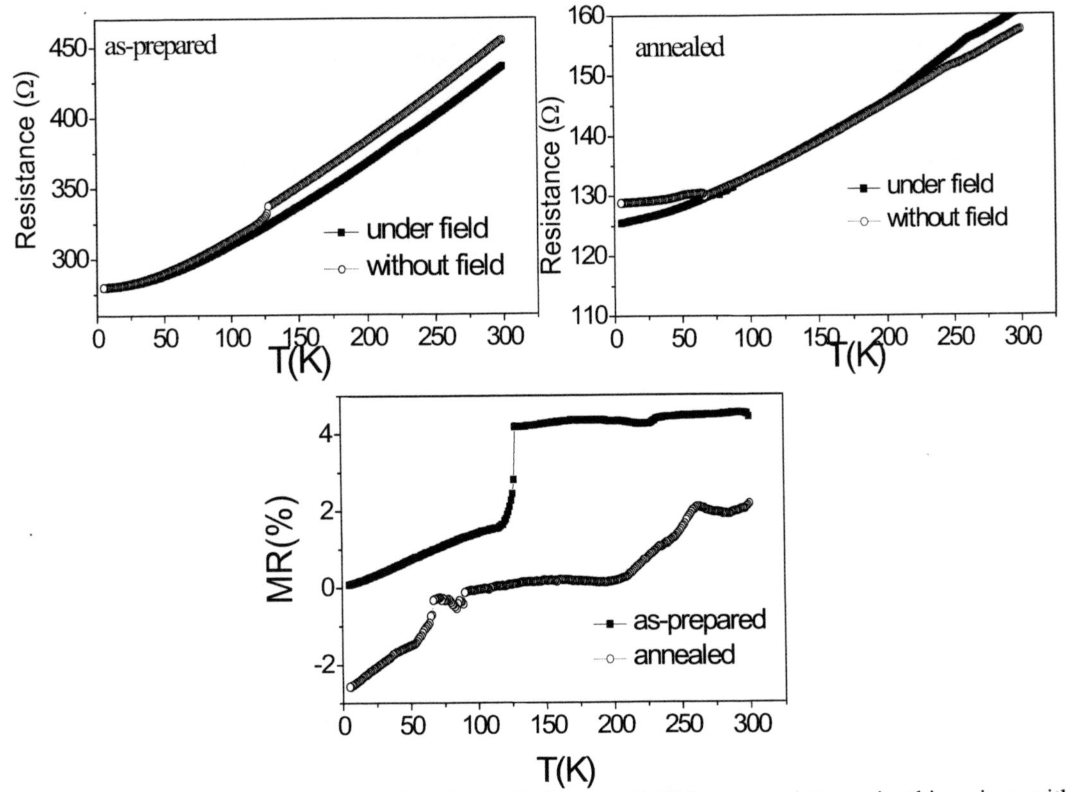

Reprinted with permission from Ref. [5] A. Zhukov et al "Magnetoresistance in thin wires with granular structure", J. Magn. and Magn., Mater. 294 (2005) 165-173. Copyright Elsevier (2005) (Figure 6).

Figure 98. Temperature dependence of MR measured in as-prepared and annealed at 973 K microwires measured at 5K -300 K.

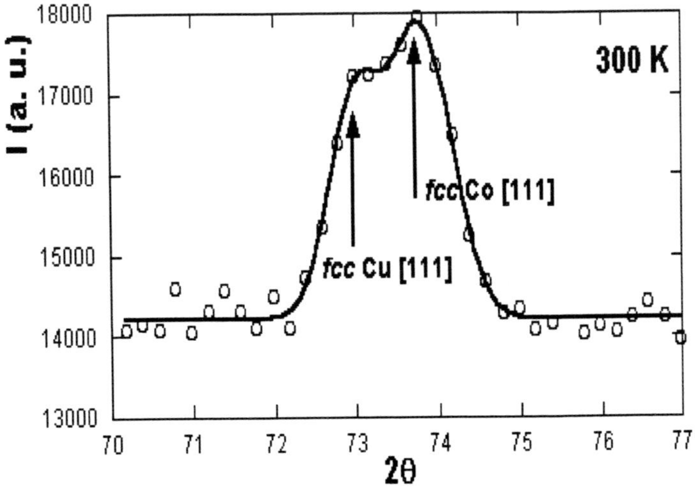

(a)

Figure 99. Continued on next page.

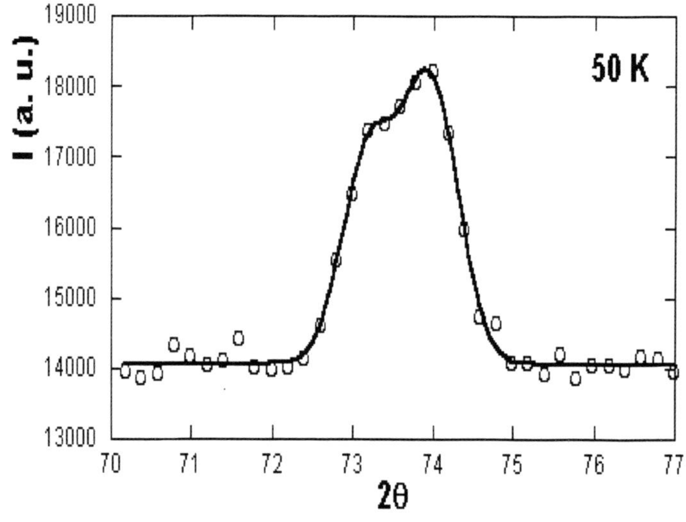

(b)

Reprinted with permission from Ref. [5] A. Zhukov et al "Magnetoresistance in thin wires with granular structure", J. Magn. and Magn., Mater. 294 (2005) 165-173. Copyright Elsevier (2005) (Figure 7).

Figure 99. Neutron diffraction spectra measured at 300 K (a) and 50 K (b).

Magneto-resistance, MR, has been measured in as-prepared $Co_{10}Cu_{90}$ microwires with metallic nucleus radius, r_m of 7 and 8.5 μm. Considerable magneto- resistance, MR, is found in both microwires (see Figure 102) in the temperature range of 5-300 K.

MR effect of the sample achieves 12% at 5K under magnetic field, H, of about 90 kOe. Magnetic field dependence of MR is completely anhysteretic showing monotonous decay with magnetic field and it is quite different from recently observed hysteretic MR effect observed in $Co_{29}Ni_{25}Mn_1Cu_{45}$ microwires [2, 5, 59]. Such shape of MR(H) dependence should be attributed to the giant magnetoresistance related with the existence of a single domain ferromagnetic particles embedded in am immiscible medium.

It is worth mentioning that MR is observed in a wide temperature range and the value of MR increases with temperature achieving at 200 K about 18%.

MR effect and granular structure also have been observed in $Cu_{63}Fe_{37}$ microwires[222, 223].

As regarding the crystalline structure, three clearly observed X-ray diffraction peaks are characteristic of the Cu phase (face centered cubic FCC, lattice parameter: 3.61 Å with atomic spacings of 2.09, 1.81 and 1.28 Å). The other two peaks are related to α-Fe phase (body centered cubic BCC, lattice parameter: 2.87 Å corresponding to atomic spacings of 2.03 and 1.43 Å). Note that the peak corresponding to 1.43 Å (α-Fe) is rather small and hardly observed. Structure consists of Cu nano-grains with the average grain size of 40 nm, roughly independent of the sample geometry (represented by the ratio ρ), as found elsewhere [222, 223]. On the other hand, the average size of Fe nanocrystals depends on the geometrical parameter varying between 6 and 45 nm, exhibiting a maximum at intermediate ρ ratio.

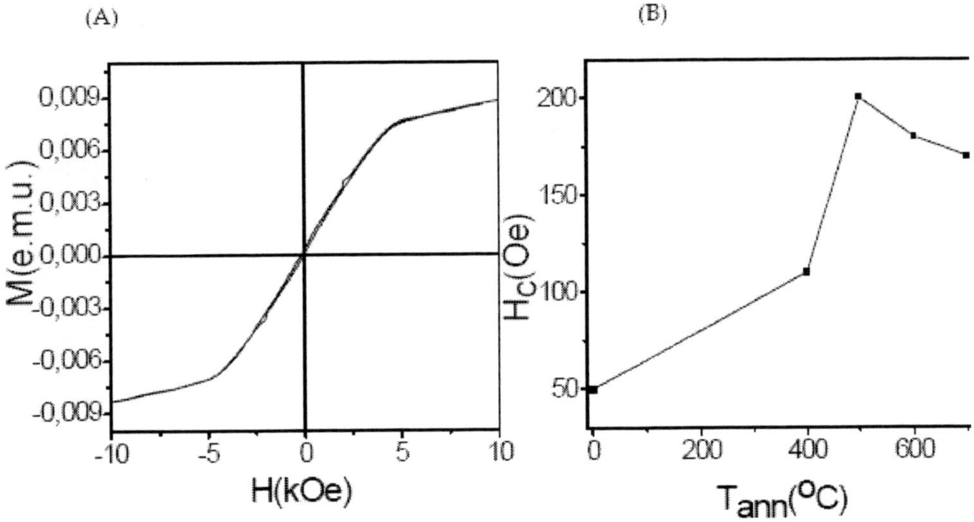

Reprinted with permission from Ref. [5] A. Zhukov et al "Magnetoresistance in thin wires with granular structure", J. Magn. and Magn., Mater. 294 (2005) 165-173. Copyright Elsevier (2005) (figure 10).

Figure 100. Hysteresis loop measured in as-prepared $Co_{10}Cu_{90}$ microwire with ρ=**0**.415.

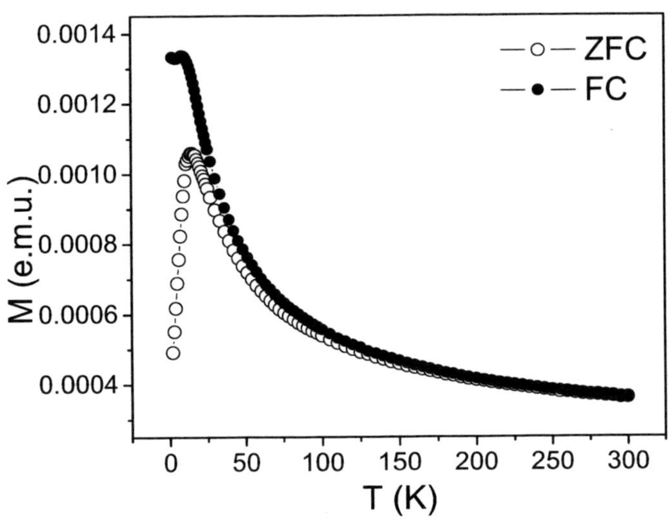

Figure 101. Zero field cooled (ZFC) and field cooled (FC) curves of $Co_{10}Cu_{90}$ microwire.

Magnetization curves measured at different temperatures are shown in Figure 103.a-b. At room temperature the coercivity is of the order of 30 Oe. Significant dependence of the hysteresis loops on the geometrical parameter has been found. All curves display both superparamagnetic and ferromagnetic features. A small hysteresis area of the measured loops and the absence of saturation indicate the presence of a superparamagnetic phase, especially significant for ρ=0.06. The ferromagnetic behaviour is evidenced by the values of remanence and coercivity, which are indeed rather low. Such behaviour can be attributed to the combined superparamagnetic-ferromagnetic phases due to co-existence of the ferromagnetic and superparamagnetic particles. Such co-existence occurs due to a wide size distribution of Fe-grain and/or concentration distribution of Fe in Cu. At the same time, temperature dependences of the saturation magnetization, $M_s(T)$, shown in the inserts of Figs. 103 (a,b), exhibit decreasing with temperature but did not present well defined Curie temperature typical for ferromagnetic materials. This probably related with Fe nanocrystals size distribution and inhomogeneity of ferromagnetic phase.

The magnetization measured under applied magnetic field (FC) and without magnetic field (ZFC) exhibit significant difference (see Figure 104). From the comparison between the two curves seen in Figs. 104 a,b it is possible to roughly determine the average blocking temperature and distribution f(TB) following a standard procedure [2, 221-224]. The obtained $f(T_B)=\frac{d(M_{ZFC}-M_{FC})}{dT}$ confirms the inhomogeneous magnetic structure assumed by analyzing the magnetization curves, consistent with previous works [222]. It follows from the analysis of Figure 104a, corresponding to $Cu_{63}Fe_{37}$ microwires with the geometrical parameter ρ=0.46, the existence of a large number of small particles with a diameter less than 30 nm and just a few large particles. On the other hand, for Figure 104b ($Cu_{63}Fe_{37}$ microwires with a geometrical parameter ρ=0.06) the estimated averaged crystal size is around 5 nm. ZFC/FC measurements performed on microwires with different geometrical parameters exhibit an intermediate behaviour. However, it is worth noting that ZFC/FC analysis in this case is only indicative, and it gives a rather rough perspective of the nanostructure. In fact, the observed ZFC/FC behaviour resembles the one expected for a rather narrow distribution of grain sizes. This apparently contradictory result was previously observed in several granular samples (see, for example, ref. 225). In fact dipolar interactions and magnetic cluster formation can lead to spurious structural data indirectly obtained from ZFC/FC curves, and thus such results are only valid when analyzed together with direct structural data, such as X-Ray diffraction and TEM analysis. Furthermore, it is worth noting that the FC curve displays an additional peak at lower temperature, indicating a more complex magnetic behaviour, probably a reentrant spin (or cluster) glass-like transition. This magnetic phase is usually present in complex granular systems, and can be related to both dipolar interactions and/or surface effects (see, for example, ref. 226).

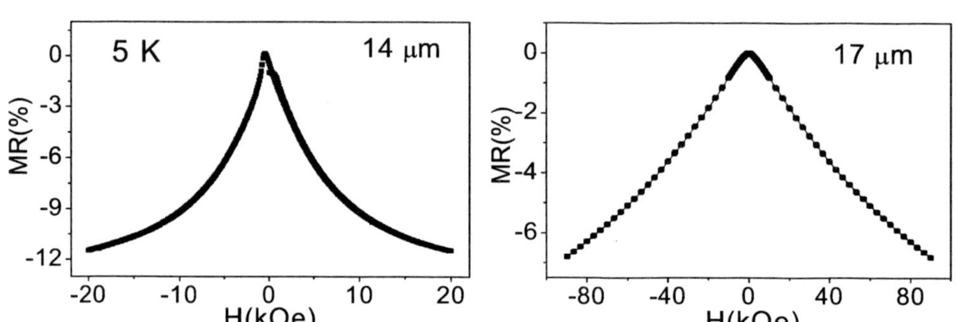

Reprinted with permission from Ref. [5] A. Zhukov et al "Magnetoresistance in thin wires with granular structure", J. Magn. and Magn., Mater. 294 (2005) 165-173 Copyright Elsevier (2005) (figure 11).

Figure 102. Magnetic field dependence of MR measured in as-prepared $Co_{10}Cu_{90}$ microwires for different geometric characteristics at 5K.

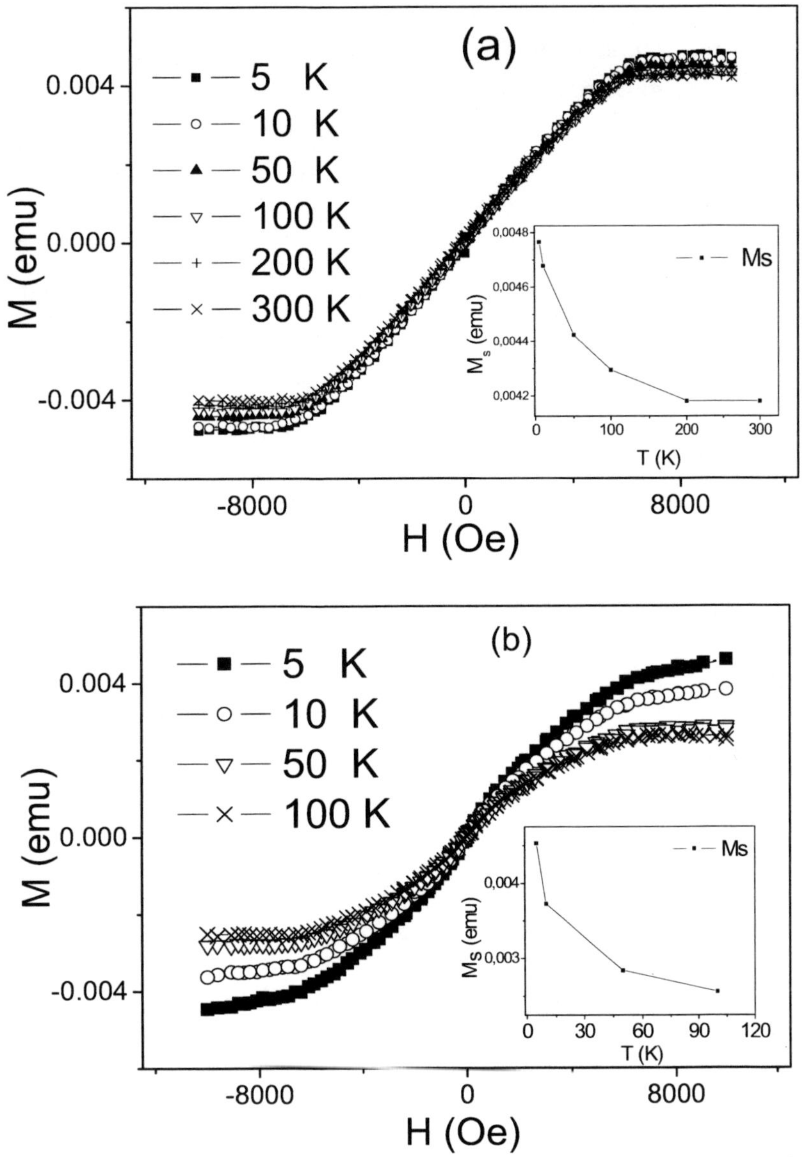

Reprinted with permission from Ref. [222] A. Zhukov et al. "Studies of Fe–Cu microwires with nanogranular structure" J. Phys.: Condens. Matter V21, 3 (2009) 035301 (6pp) Copyright © IOP PUBLISHING(2009) (Figure 3).

Figure 103. Hysteresis loop for the $Cu_{63}Fe_{37}$ as cast sample at different temperatures with different ratio ρ values (a) 0.46 (b) 0.063. Evolution of magnetization at saturation is shown at the inset.

The behaviour of the main magnetic phase (α-Fe) is very important in this type of materials. Note that the content of α-Fe phase and the average grain size is lower for the sample $Cu_{73}Fe_{37}$ with ρ=0.063. The obtained content of α-Fe phase varies around 1% for ρ=0.06, 8% for ρ=0.16 and ρ=0.31, 7% for ρ=0.46, as can be found elsewhere [227]. It should be noted, that following the previously proposed scheme for these type of samples [228], the loops show a superparamagnetic character for ρ=0.063 and a ferromagnetic one for ρ=0.46. This ferromagnetic character could be related to the aforementioned higher content of α-Fe phase and larger average grain size for this sample and, in a certain way, there is a correlation between magnetic measurements and structural results.

It should be noted also that the field at which the saturation is roughly achieved is quite large (around 6 kOe) (see Figs.103). But complete saturation of the sample with lowest ρ–ratio is not reached even at this magnetic field (Figure 103b).

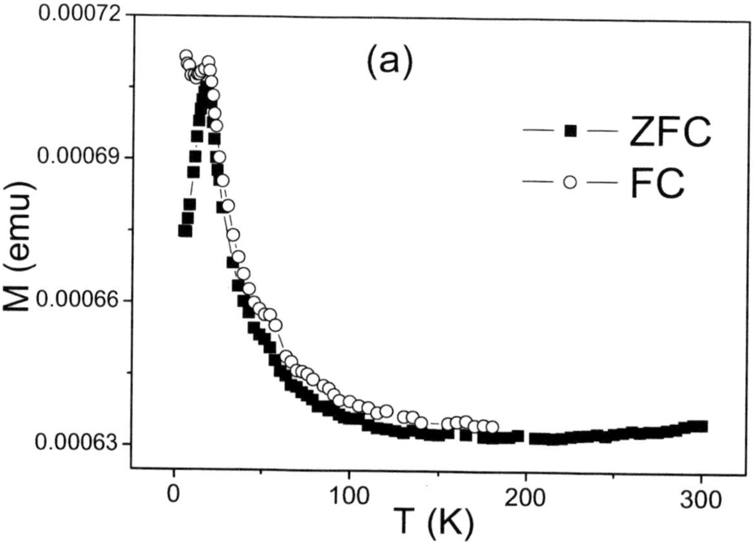

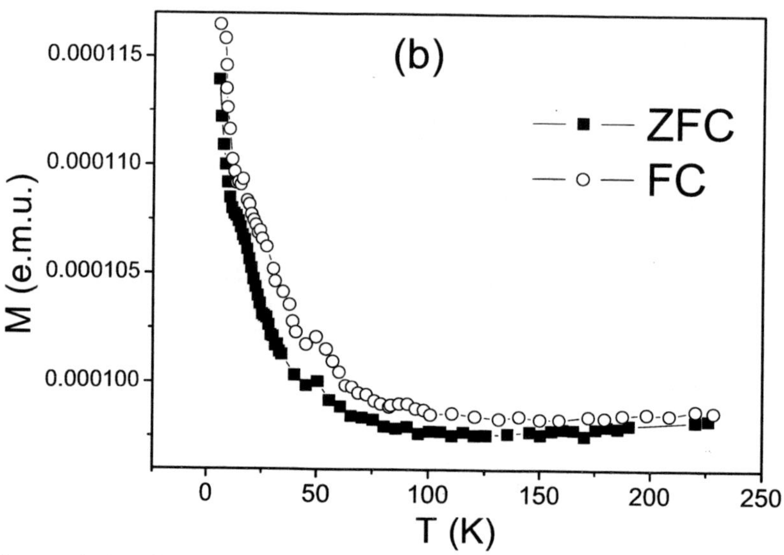

Reprinted with permission from Ref. [222] A. Zhukov et al. "Studies of Fe–Cu microwires with nanogranular structure" J. Phys.: Condens. Matter V21, 3 (2009) 035301 (6pp) Copyright © IOP PUBLISHING(2009) (Figure 4).

Figure 104. Zero field cooled and field cooled curves for the as-cast $Cu_{63}Fe_{37}$, measured at H_{DC}=100 Oe up to 300 K with different ratio ρ values (a) 0.46 (b) 0.063.

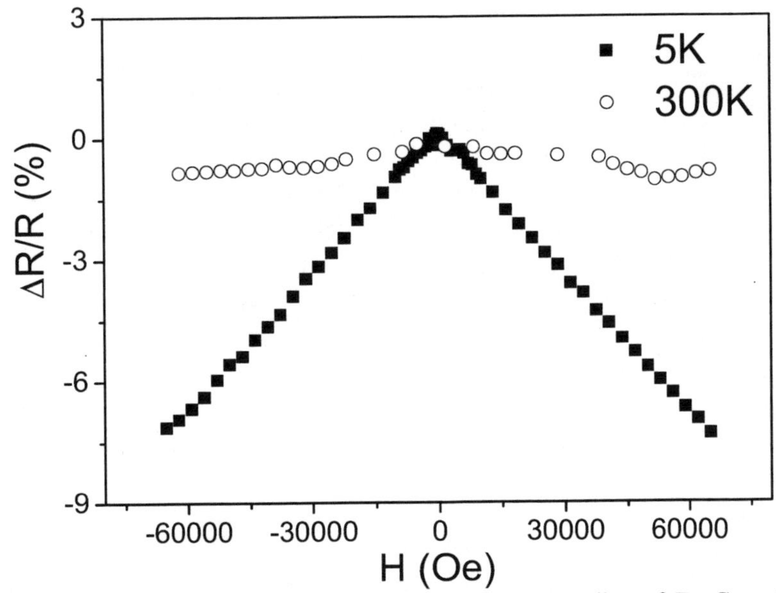

Reprinted with permission from Ref. [222] A. Zhukov et al. "Studies of Fe–Cu microwires with nanogranular structure" J. Phys.: Condens. Matter V21, 3 (2009) 035301 (6pp) Copyright © IOP PUBLISHING(2009) (Figure 5).

Figure 105. Magnetoresistance of the as-cast $Cu_{63}Fe_{37}$ microwire with ρ=0.46 measured at 5 K and 300K at magnetic field from -7 to 7 T.

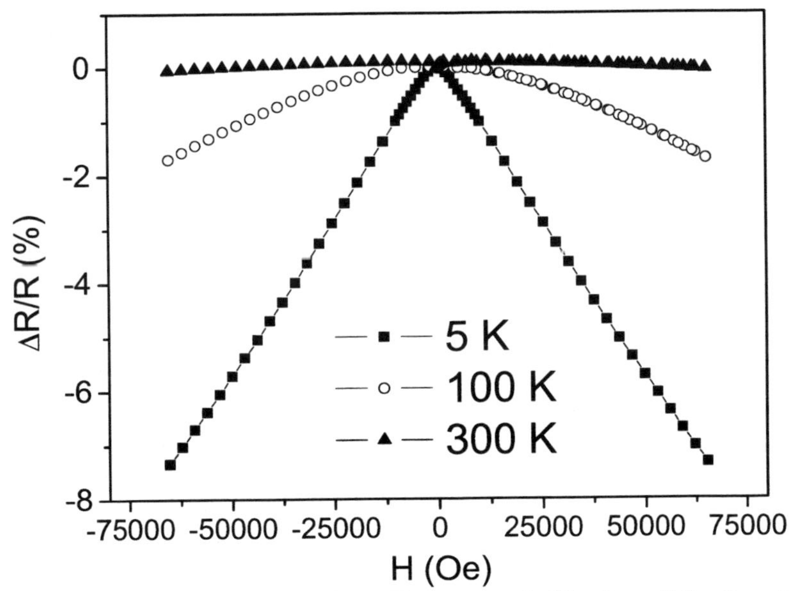

Reprinted with permission from Ref. [222] A. Zhukov et al. "Studies of Fe–Cu microwires with nanogranular structure" J. Phys.: Condens. Matter V21, 3 (2009) 035301 (6pp) Copyright © IOP PUBLISHING(2009) (Figure 6).

Figure 106. Magnetoresistance of the as-cast $Cu_{63}Fe_{37}$microwire with ρ=0.31 measured at 5 K, 100K and 300K at magnetic field from -7 to 7 T.

One of the possible explanations is that the axis of the microwires is the hard direction. But this assumption contradicts to the logics, since the demagnetizing factor of the sample with lowest ρ– value when applied magnetic filed is along the microwire axis takes lowest values. Additionally, from the experimental data on Fe nanowires it is known that usually Fe nanowires have easy axis along the nanowire axis [228]. It was assumed also that higher contribution of superparamagnetic behaviour of the thinnest microwire related also with the dependence of the structure on ρ–ratio can be attributed to the higher internal stresses induced by the difference in thermal expansion coefficients of simultaneously solidifying glass and metal [2, 224]. In fact it is well established that the strength of the internal stresses, σ, is related to the glass coating thickness through the difference in the thermal expansion coefficients of metallic nucleus and outer glass coating solidifying simultaneously. The estimated values of the internal stresses in these glass coated microwires arising from the difference in the thermal expansion coefficients of metallic nucleus and glass coating are of the order of 100-1000 MPa, depending strongly on the ratio between the glass coating thickness and metallic core diameter [2-4], increasing with the increasing of the glass coating thickness. These internal stresses can affect the grain size of Fe grains. Similar behaviour has been observed in Finemet –type microwires when the nanocrystallization process has been affected by the ρ– ratio [209-211]. To explain this, it was suggested [2, 209-211], that the internal stresses in the metallic nucleus hinder the segregation of crystalline phases.

The MR effect in the range of 7 % has been also observed for ρ=0.46 and ρ=0.31 $Cu_{73}Fe_{37}$ glass coated microwires (Figs 105 and 106 respectively). The shape of magnetic field dependence is typical for the GMR effect, when resistance decreases with increasing magnetic field. Decreasing the temperature MR increases (from 0.05% (300 K) till 7.6% (5 K) for the sample with ρ=0.31 and from 1% till 7 % for the sample with ρ=0.46). In most systems exhibiting GMR, the GMR ratio increases decreasing the temperature [230-231]. Usually this increase is attributed to the superparamagnetic origin of grains responsible for the GMR behaviour as well as related with temperature dependence of resistivity [230,231].

The field dependent resistivity in granular materials is related to a spin-dependent scattering of conduction electrons within the magnetic particles as well as the interfaces between magnetic and non magnetic regions (however, it is assumed that the interfacial effect contributes to a dominant extent to GMR [230-232]). Because of the complex structure of the granular materials, the relationship between microstructure and GMR is still not fully understood.

In the case of glass—coated microwires the internal stresses induced mostly by the difference of thermal expansion coefficients of glass and metal are quite significant [2-4]. Therefore studies of abovementioned composition with different geometry can be interesting in order to reveal the internal stress contribution in magnetic and magneto-transport properties. Similar studies in Co-Cu microwires confirmed that indeed crystalline structure and GMR effect is affected by the glass coating thickness and metallic nucleus diameter [219]. These internal stresses can affect the structure and magnetism of small Fe grains inside the Cu-matrix, as has been previously observed for the case of epitaxially grown Fe films on Cu(100), where bcc- and fcc - like structures were observed depending on the film thickness [232]. These studies in Cu-Fe microwires are in progress.

Concluding, considerable (up to 18 %) magnetoresistance has been observed in glass coated $Co_{10}Cu_{90}$, $Cu_{63}Fe_{37}$ and $Co_{29}Ni_{25}Mn_{1}Cu_{45}$ microwires prepared out of immiscible elements by the modified Taylor-Ulitovsky method. First structure, magnetic and magneto-transport characterization have been performed.

Magnetic and transport properties of $Co_{29}Ni_{25}Mn_1Cu_{45}$ microwires depend on the annealing. Thus the coercivity in as-prepared state is ranging of order of tens of Oe and can be enhanced by thermal treatments reaching values of up to 750 Oe. Magnetic field dependence of MR in as-prepared and annealed samples is quite different. Roughly monotonic decay with magnetic field is observed in as-prepared microwires. MR effect diminishes decreasing the temperature. Annealed sample exhibits a maximum and considerable hysteresis on magnetic field dependence of MR phenomenon. The position of this maximum correlates well to the values of H_c obtained from magnetization curves.

Magnetic and transport properties of $Co_{10}Cu_{90}$ microwires depend on the sample geometry. Considerable MR (up to 18%) has been observed in $Co_{10}Cu_{90}$ microwires.

$Cu_{63}Fe_{37}$ microwires structure consists of Cu nanograins with average grain size of around 40 nm and α-Fe nanocrystals with average grain size ranging between 6 and 45 nm depending on samples geometry. The resulting content of α-Fe phases also strongly depends on the sample geometry. Experimental data on magnetic and magneto-transport behaviour are interpreted assuming the mixed ferromagnetic-superparamagnetic behaviour of the fabricated samples. Significant Magnetoresistance (about 7%) has been found in studied samples. The shape of observed dependences is typical for GMR effect.

5.2. Magneto-Caloric Effect in Glass-Coated Microwires

Ferromagnetic glass-coated microwires have a number of attractive features that make them strong candidates for use as a sensing element in high performance magnetic field or stress sensors, magnetic labels, and micro machines [2, 119]. Although the research on magnetic microwires is well established, it was only very recently proposed to utilize them within artificially structured materials. As a new example, magnetic glass-coated microwires of Fe_3P and SnMn composition have been proposed as magnetocaloric materials to engineer magnetic functionality of advanced materials, for example, for exhibiting significant magnetic entropy changes around the Curie temperature (T_C) associated to MCE due to magnetic wire actuating performance [233]. Therefore, a type of a smart material is proposed exhibiting MCE and coated with chemically inactive layer. For this purpose two Fe_3P and SnMn thin glass coated microwires with different total diameters of 14-34 μm were produced by the Taylor-Ulitovski method [233]. The master alloy of the composition was prepared by arc melting of the pure elements in Ar atmosphere. Subsequently, when the metallic alloy and the Pyrex glass coating were simultaneously molten, the so-formed microwire was drawn and rolled onto a rotating cylinder and quenched to room temperature. The samples obtained were in the form of a tiny metallic wire with the dimensions above mentioned.

Thermal dependence of the ac-magnetic susceptibility of the different glass coated microwires studied in this work has been determined from 78 K to 400 K. Adiabatic temperature change, originated from MCE has been measured directly on the special set-up created in Department of Physics of MSU. The thermo-couple was utilized for continuous registration of the temperature on the magnetic field change from 0 kOe to 12 kOe at certain temperatures. During the measurements sample has been placed in the chamber evacuated up to the 10^{-3} Torr to maintain the adiabatic conditions. The estimated error due to the heat leakage was found to be of $5*10^{-3}$ K.

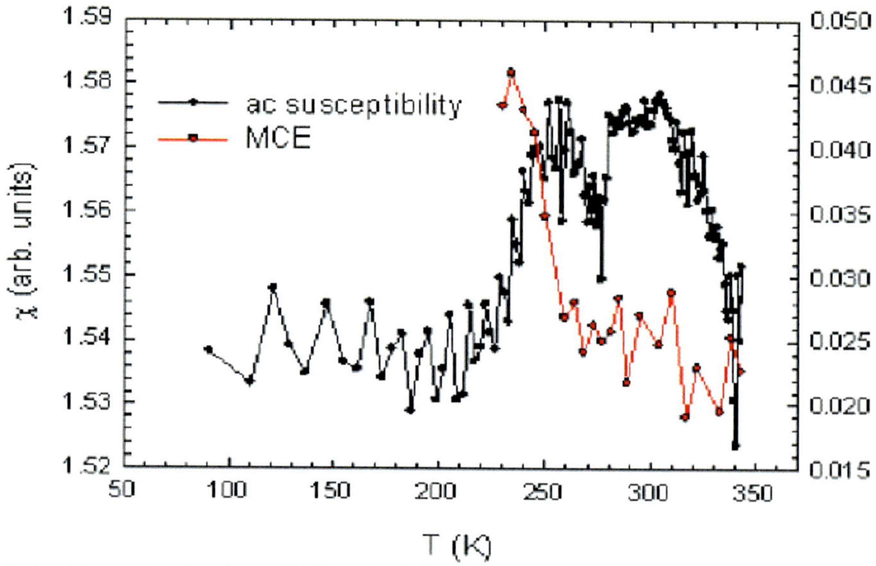

Reprinted with permission from Ref. [233] M. I. Ilyn et al. "Magnetocaloric effect in nanogranular glass coated microwires" Phys. Stat. Sol. (a), 205 No 6 (2008) 1378-1381 Copyright © 2008 WILEY-VCH Verlag GmbH & Co. KGaA, Weinheim (Figure 2).

Figure 107. Ac-susceptibility and MCE in the SnMn as a function of the temperature.

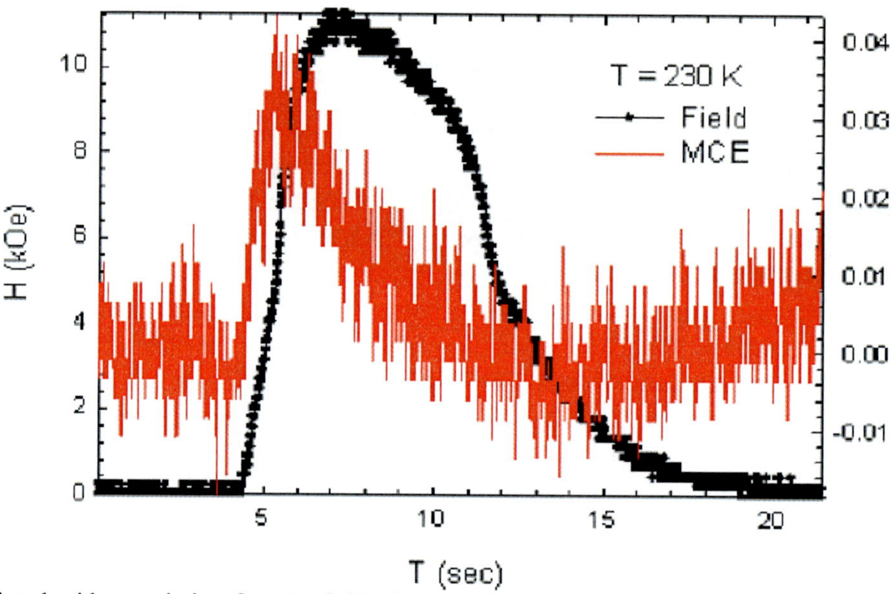

Reprinted with permission from Ref. [233] M. I. Ilyn et al. "Magnetocaloric effect in nanogranular glass coated microwires" Phys. Stat. Sol. (a), 205 No 6 (2008) 1378-1381 Copyright © 2008 WILEY-VCH Verlag GmbH & Co. KGaA, Weinheim (Figure 3).

Figure 108. Temporal dependence of the MCE in SnMn.

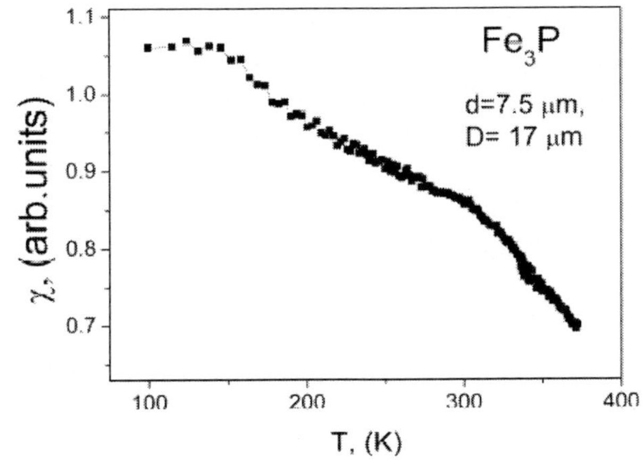

(a)

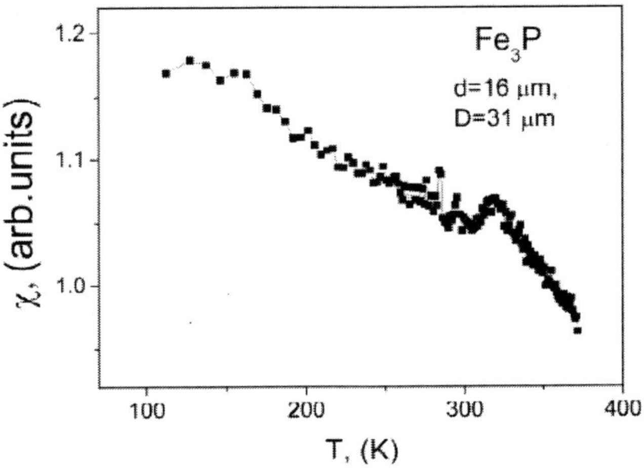

(b)

Reprinted with permission from Ref. [233] M. I. Ilyn et al. "Magnetocaloric effect in nanogranular glass coated microwires" Phys. Stat. Sol. (a), 205 No 6 (2008) 1378-1381 Copyright © 2008 WILEY-VCH Verlag GmbH & Co. KGaA, Weinheim (Figure 4).

Figure 109. Ac-susceptibility of Fe_3P with different geometry.

The temperature dependence of ac-susceptibility of SnMn microwire is shown in the Figure 107. The second order phase transition from ferromagnetic to paramagnetic state is accompanied with the peak of the ac-susceptibility. However in the ordered state temperature dependence of the ac-susceptibility is a complex function of the magnetization saturation, anisotropy and coercive field. The nonmonotonic variation of the susceptibility of SnMn implies that the characteristic peak near the phase transition is overlapped with the peak originated from the peculiarities of the domain structure and can not be properly distinguished. Nevertheless in the paramagnetic state susceptibility decreases according to Curie-Weiss low and the inflection point on the temperature dependence of the ac-susceptibility can be used as an estimation of the place of the phase transition. The Curie temperature of the sample, defined as a temperature of the maximum $|d\chi \backslash dT|$, was found to be of ~340 K.

A plot of temporal dependence of the adiabatic temperature change collected on the field change of 12 kOe is shown on the Figure 108. Application of the magnetic field gives rise to the ΔTad of 0.04 K. A set of field dependences of the ΔTad was utilized to produce a curve of the temperature dependence of MCE presented on the figure 108. The largest value of ΔTad was observed well below the Curie temperature implying that the transformation of the domain structure and rotation of the magnetic moment due to the magnetization in SnMn give rise to the larger values of the MCE than the change of the magnetization saturation itself which is proportional to |dI\dT| and reaches the maximum at the Curie temperature [234].

The temperature dependences of the ac-susceptibility of Fe_3P microwires are shown in the figures 109 a,b respectively. Both curves are consisted of peaks placed slightly below the Curie temperature of ~360 K, followed by the gradual rise of the susceptibility with decreasing of the temperature. The observed elevation of the χ on the cooling probably originated from the complex domain structure of the samples. The change of diameters doesn't shift the Curie temperatures but affects on the dynamics of magnetization giving rise to the difference in the cusps of the χ. The MCE in the both samples were found to be less than in SnMn attaining the largest value of 0.02 K in the vicinity of the Curie temperature.

Concluding, measurements of the ac-susceptibility of three glass coated metallic microwires have been performed for fast screening of their magnetic properties. The subsequent direct measurements of the adiabatic temperature change originated from the MCE have been done in the range of the temperatures covering the locations of the extremum of ac-susceptibility. The observed values of ΔTad are rather small being of ~0.04 K in the SnMn and ~0.02 K in both Fe_3P samples. However it is the first direct measurements of the MCE in this kind of objects. The magnetic glass coated microwires are flexible substances readily available in the wide range of chemical compositions capable of easy variation of their properties. Hopefully these investigations will become a new direction in the development of the magnetocaloric materials.

Chapter 6

FREQUENCY DEPENDENCE OF COERCIVITY. EFFECT OF MAGNETIC STRUCTURE

The coercivity is commonly considered to be one of the most important parameter in applied magnetism [200, 201, 235, 236]. Coercivity, H_c, is associated with the pinning well of the domain wall. Frequency, f, dependence of the coercivity, $H_c(f)$ in different magnetic materials has been observed [235-237]. It was found that depending on domain structure of magnetic materials affects $H_c(f)$ dependence and can give an additional information on magnetic structure of these magnetic materials.

Previously frequency dependence of the coercivity was measured and analyzed in magnetic wires or ribbons [200, 235]. Thus it was shown, that the coercivity in most magnetic materials at low enough frequency f can be phenomenologically expressed as:

$$H = H_{c0} + const(fH_0)^{1/n}, \quad (56)$$

where H_{c0} is the static coercivity, H_0 is the applied field amplitude and n is the coefficient ranging from 1 to 4 depending on sample geometry and the type of the hysteresis loop of the studied materials.

On the other hand the dependence of H_c on the frequency was obtained from the solution of the equation of the domain wall motion for few particular cases in [200, 201, 235]. It should be stated that the frequency dependence of coercivity has been analyzed for the case of magnetically bistable samples when the remagnetization takes place as the depinning of existing domain wall and consequent domain wall propagation. In this case below some critical magnetic field associated with the single and large Barkhausen jump there is only a reversible magnetization process associated with reversible domain wall movement. Above such critical field large and single Barkhausen jump occurs. Such peculiar behaviour has been observed particularly in amorphous wires with high enough magnetostriction constant and Fe-rich glass-coated microwires. It is reasonable to associate such critical magnetic field with a critical displacement of the domain wall, taking into account that above such critical displacement the irreversible domain wall propagation takes place [200].

Experimental studies show that the coefficient n can take values ranging from 1 to 4 depending on sample geometry and the type of the hysteresis loop of the studied materials, changing even for the same material after different thermal treatments [235].

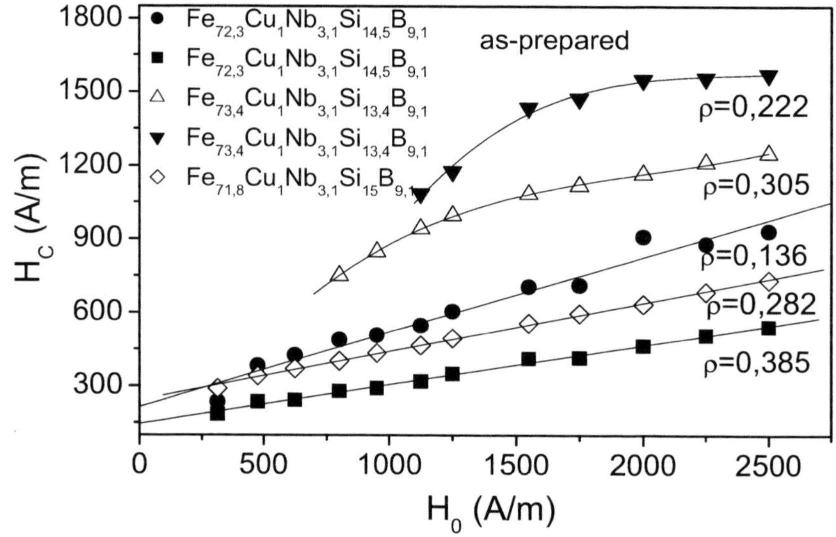

(a)

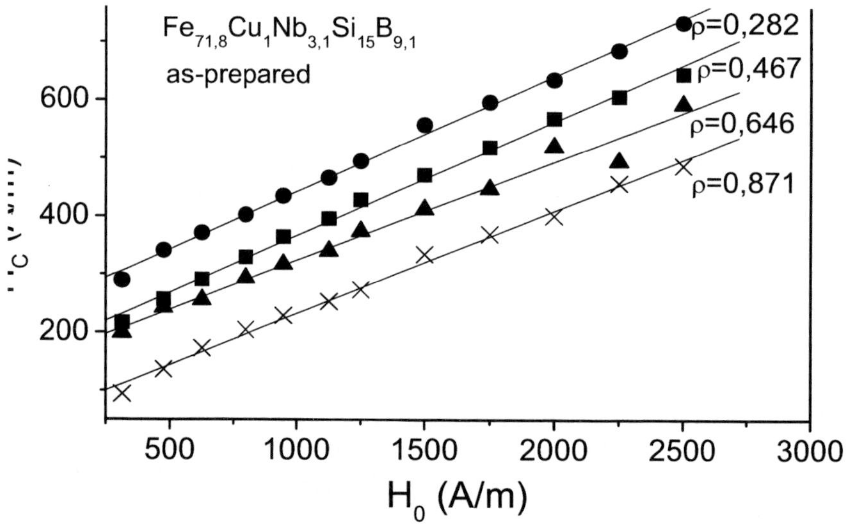

(b)

Reprinted with permission from Ref. [209] V. Zhukova et al., NanoStructured Materials Vol.11, No.8, 1319-1327 (1999). Copyright Elsevier Science Ltd Acta Metallurgica Inc. (2000) (Figure 5).

Figure 110. Dependence of coercivity on the amplitude of the applied magnetic field for different microwires in as-prepared state $Fe_{73.4-x}Cu_1Nb_{3.1}Si_{13.4+x}B_{9.1}$, (A) x = 0; 1.1; 1.6; (B) x = 1.6.

Analysis of frequency and/or magnetic field amplitude dependence of coercivity have been performed in few papers. Considering the external magnetic field $H(t) = 4fH_o$ (f- magnetic field frequency, H_o – magnetic field amplitude) there are various particular solutions of the domain wall motion equation [200, 201, 235-236]:

$$Md^2x/dt^2 + Ldx/dt + Kx = 2I_sH_i \quad (57)$$

where M is the effective mass of domain wall, L- damping coefficient, K-elastic coefficient, x- domain wall displacement, $H_i=H+H_{ec}$, H_i – intrinsic magnetic field, H – external field, H_{ec} – the field created by local eddy currents. In fist approximation can be assumed [200] that $H_{ec}=Adx/dt$, where A is the coefficient.

This equation has been re-written [194, 195, 229] as:

$$d^2x/dt^2 + pdx/dt + rx = 1 \quad (58)$$

where $p= (r+2I_s A)/M$, $r=K/M$ and $l=2I_sH/M$

There are several solutions of (3) for different simplified cases [200,201]:

1) High frequency (or small time)

$$H_c= 4fH_0t= 4fH_0 (\alpha - (\alpha-[4\alpha^4x_{cr}]^{1/2}/(1-2\alpha^4)^{1/4}) \sim fH_0 \quad (59)$$

where – x_{cr} critical displacement of the domain wall to be depinned i.e. related with the critical magnetic field, $\alpha =1/p$

2) Low frequency, large times,

$$H_c=H_{c0}+ 4 fH_0(L+2I_sA)/K\sim H_{c0}+ BfH_0 \quad (60)$$

where is the proportionality constant

3) Small M, i.e. neglecting first term of eq. (55).

3a) High frequency or small times

$$H_c=H_{c0}/3 + (4 fH_0)^{1/}3[4 x_{cr} +2I_sA)/ I_s]^{1/2} \quad (61)$$

3b) Low frequency, large times

$$H_c=3H_{c0} + fH_0\, 8I_sA/ K \quad (62)$$

4) Small K (K<< M d2x/dt2), i.e. the case of depinned domain wall

4a) Low times, high frequencies

$$H_c =2(3Mx_{cr}/I_s)(4fH_0)^{2/3} \sim(fH_0)^{2/3} \quad (63)$$

4b) Low frequencies, large time

$$H_c=H_{c0}+4A\ (I_sx_{cr}/L+2I_sA)^{1/2}(fH_0)^{1/2} \quad (64)$$

In terms of the interpretation of aforementioned domain wall propagation equation [200,201], few kinds of experimentally observed $H_c(f)$ dependences can be interpreted considering depinned domain walls, i.e. neglecting elastic coefficient K of the domain wall motion equation. The $H_c \sim \sqrt{f}$ corresponds better to low frequency limits, i.e. when the relaxation time of the domain wall is small, while $H_c \sim f^{2/3}$ is attributed to the high frequency behaviour when the relaxation time of the domain wall is high.

From point of view of the model, magnetic amplitude and frequency dependence of coercivity has the same origin.

$H_c(H_o)$ dependences have been measured in $Fe_{73.4-x}Cu_1Nb_{3.1}Si_{13.4+x}B_{9.1}$, (x = 0; 1.1; 1.6) microwires (see Figure 110 in as-prepared state). Most of samples showed linear increase of H_c with H_o [209,238].

Figure 111 (a, b and c) shows the dependence of coercivity with the magnetic field amplitude, H_o, with the annealing temperature as parameter. For values of d/D = 0.282, 0,467 and 0.871 respectively a quite linear increase of H_c with H_o was obtained except for the sample treated at high temperature (923 K sample with d/D = 0.871 and 973 K for d/D = 0.282) where this increase of $H_c(H_o)$ results to be more significant. Regarding the H_c (H_o) behaviour found in these glass covered microwires, a linear behaviour of $H_c(H_o)$ curves are observed in those microwires exhibiting bistable behaviour [209, 238]. In ref. [200] a similar study on the effect of the amplitude and frequency of the applied magnetic field on the coercive field in amorphous alloys (ribbons, wires and microwires) was performed. Almost linear dependence of H_c with the amplitude and frequency for microwires with bistable behaviour has been observed in the case of as-prepared microwires with rectangular hysteresis loops. To explain this linear dependence a nucleation mechanism was assumed (as responsible of the bistability) with a large value of the anisotropy constant, K, for the inner core of the metallic nucleus. Consequently, deviations from linear dependences observed in some cases for as-prepared microwires and also observed in some of annealed microwires should be associated with partial crystallization of as-prepared microwires and with destroying of magnetic bistability due to re-crystallization of amorphous precursor by annealing.

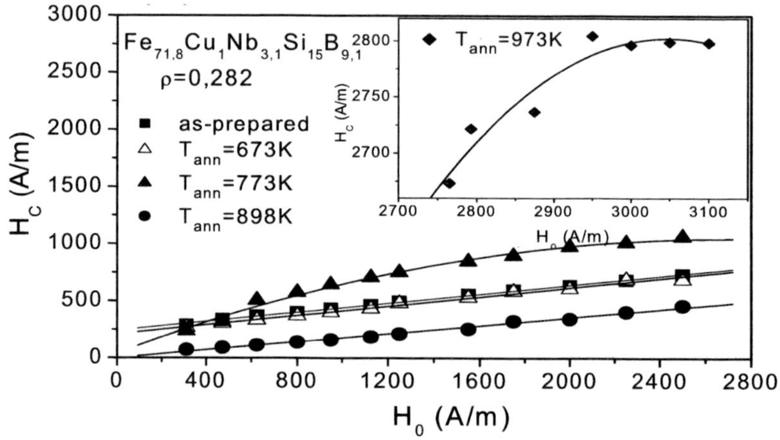

(a)

Figure 111. Continued on next page.

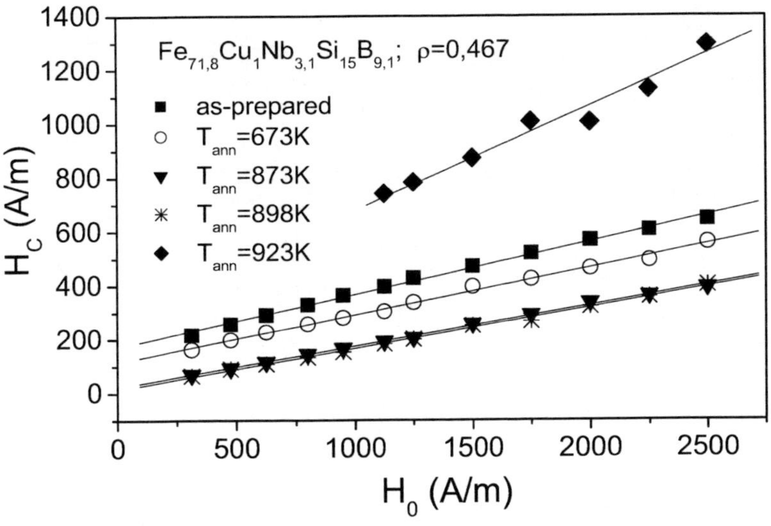

(b)

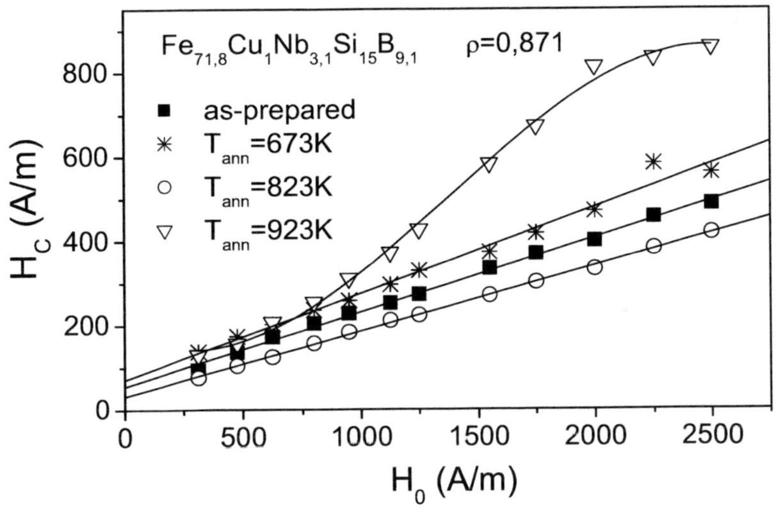

(c)

Reprinted with permission from Ref. [ ] V. Zhukova et al., NanoStructured Materials Vol.11, No.8, 1319-1327, 1999, Copyright Elsevier Science Ltd Acta Metallurgica Inc. (2000) (Figure 6).

Figure 111. Dependence of coercivity on the amplitude of the applied magnetic field for microwires $Fe_{71.8}Cu_1Nb_{3.1}Si_{15}B_{9.1}$ (A) $d/D = 0.282$; (B) $d/D = 0.467$; (C) $d/D = 0.871$.

The significant difference of the $H_c(H_o)$ curves of the microwires treated at high temperatures (above 873 K) or those without bistability in as-cast state could be connected mainly to pinning and propagation mechanisms for the coercivity. This assumption is in agreement with the hysteretic behaviour and should be associated magnetic hardening of annealed sample, although this magnetic hardening results to be quite different depending on the *d/D* ratio values and, consequently, the internal stresses on the metallic nucleus due to the glass coating thickness.

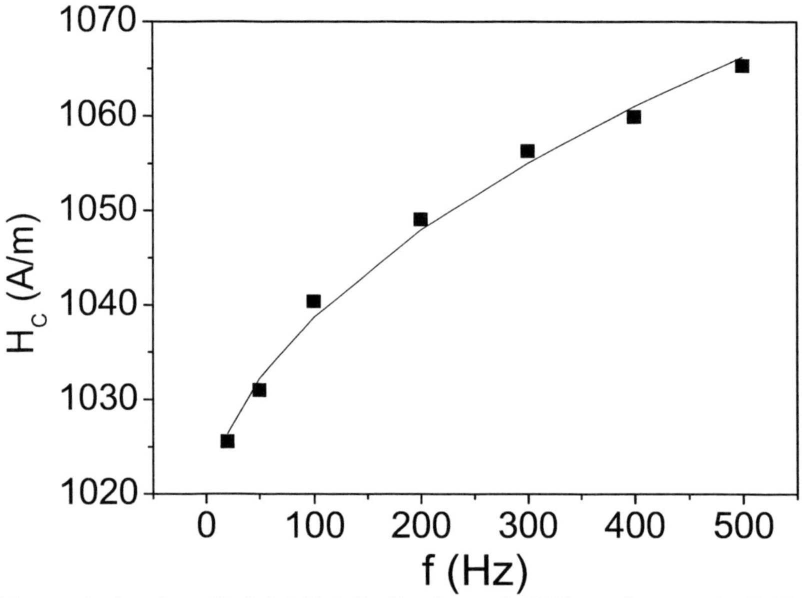

Reprinted with permission from Ref. [201] C. García et al. "Effect of magnetic field frequency on coercivity behavior of nanocrystalline Fe79Hf7B12Si2 glass-coated microwires", Physica B 403 (2008) 286–288, Copyright Elsevier (2008) (Figure 1).

Figure 112. Frequency dependence of coercive field for as-prepared $Fe_{79}Hf_7B_{12}Si_2$ samples.

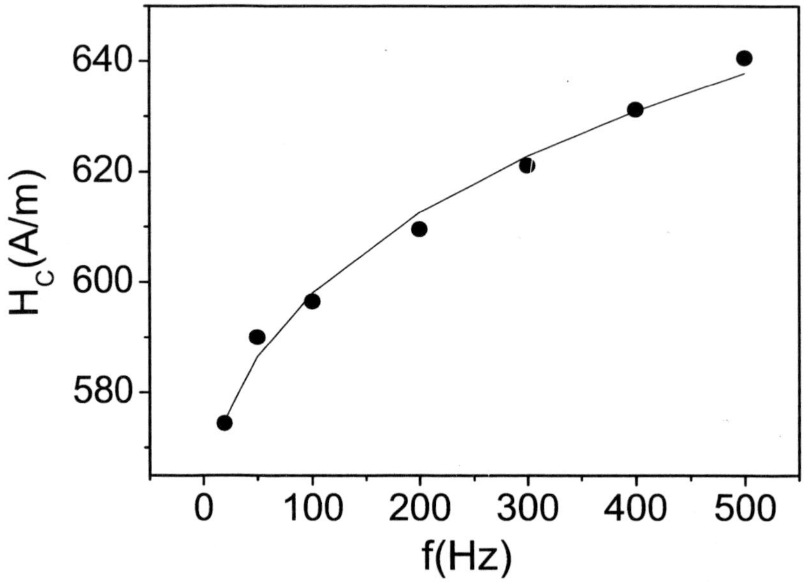

Reprinted with permission from Ref. [201] C. García et al. "Effect of magnetic field frequency on coercivity behavior of nanocrystalline Fe79Hf7B12Si2 glass-coated microwires", Physica B 403 (2008) 286–288, Copyright Elsevier (2008) (Figure 2).

Figure 113. Frequency dependence of coercive field of $Fe_{79}Hf_7B_{12}Si_2$ samples after annealing at 823K.

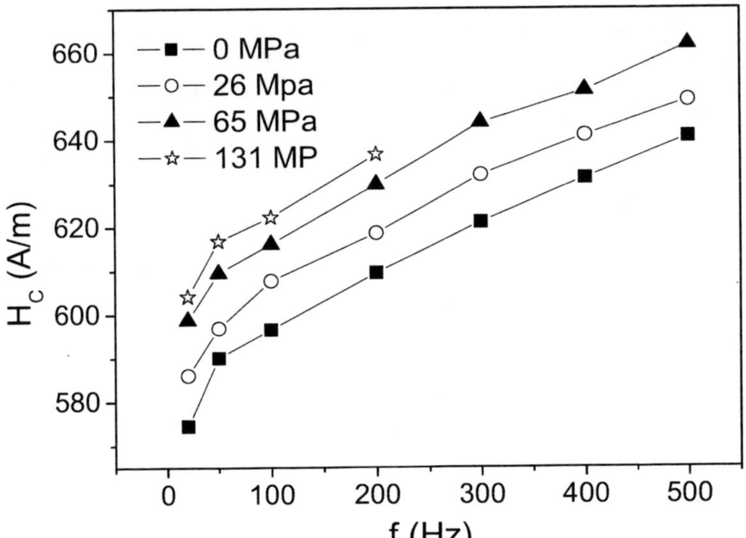

Reprinted with permission from Ref. [201] C. García et al. "Effect of magnetic field frequency on coercivity behavior of nanocrystalline $Fe_{79}Hf_7B_{12}Si_2$ glass-coated microwires", Physica B 403 (2008) 286–288, Copyright Elsevier (2008) (Figure 3).

Figure 114. Effect of applied stress on H_c *(f)* dependence.

Recently $H_c(f)$ dependence in novel glass-coated $Fe_{79}Hf_7B_{12}Si_2$ microwires with nanocrystalline structure where the structure (average grain size and the volume of crystalline phase) depend on the annealing conditions (time and temperature).

Dependence of coercivity, H_c, on magnetic field frequency, f is shown in Figure 112. Coercivity exhibits an increase with magnetic field frequency. For the as-prepared microwires the frequency dependence fits well to $\sqrt{f}$. In annealed at 550°C samples $H_c \sim f^{2/3}$ dependence is observed (see Figure 113).

In terms of the interpretation of aforementioned domain wall propagation equation [200], both kinds of experimentally observed $H_c(f)$ dependences (see Figs. 112, 113) can be interpreted considering depinned domain walls, i.e. neglecting elastic coefficient K of the domain wall motion equation. The $H_c \sim \sqrt{f}$ corresponds better to low frequency limits, i.e. when the relaxation time of the domain wall is small, while $H_c \sim f^{2/3}$ is attributed to the high frequency behaviour when the relaxation time of the domain wall is high.

Applied stress does not change significantly the character of $H_c(f)$ dependence, although the of the frequency independent contribution H_{c0} increases (see Figure 114).

It should be noted that in some cases non-monotonic $H_c(f)$ dependences have been observed [239, 240] at low frequencies. Additional efforts are needed to understand better all aspects of $H_c(f)$ dependences in magnetic materials.

Summarizing, the frequency dependence of the coercivity in amorphous and nanocrystalline magnetic microwires is observed and explained in terms the solution of the equation of the domain wall motion. For amorphous and Finemet-type nanocrystalline microwires linear dependence of H_c, on magnetic field frequency, f (or amplitude H_0) has been observed. For novel nanocrystalline microwires of $Fe_{79}Hf_7B_{12}Si_2$ composition in as-prepared sample fits to $H_c \sim \sqrt{f}$ corresponding better to low frequency limits, i.e. when the relaxation time of the domain wall are small, while the annealed sample exhibit $H_c \sim f^{2/3}$ which can attributed to the high frequency behaviour when the relaxation time of the domain wall are high and considering depinned domain walls, i.e. neglecting elastic coefficient K of the domain wall motion equation.

Chapter 7

APPLICATIONS

Existing and future applications of glass-coated microwires are determined by magnetic and magneto-transport properties. Consequently, applications of magnetic microwires can be classified as following:

1. Magnetic sensors and devices based on GMI effect
2. Magnetic sensors and devices based magnetic bistability and domain wall propagation
3. Magnetometers using almost zero magnetostrictive amorphous and nanocrystalline materials (flux sensors, current sensors, proximity sensors, and magnetometers)
4. Stress sensors using high magnetostriction constant (λ_s = 2 - 4.5 $x10^{-5}$). These include:
5. Thermal sensors utilizing the Curie temperature (T =-50 ÷ 250 °C)
6. Sensors based on magnetoelectric effects (inverse Wiedemann ad Matteucci)
7. Stress- tuneable and temperature-tuneable composite materials based on thin ferromagnetic wires

At this moment the industrial application of magnetic microwires is related with high and linear response of the off-diagonal GMI effect achieved in cold-drawn microwires [7, 161] and recently also observed in thin glass-coated microwires [4, 57, 180, 241,242].

Additionally a number of magnetic sensors based on giant magneto-impedance (GMI) effect and stress-impedance (SI) effect with the C-MOS IC circuitry and advantageous features comparing with conventional magnetic sensors have been reported [8, 159-161, 243]. Main possible applications reported in Ref. [8, 159-161, 243] are related with the detection of low magnetic fields, small weights and vibrations in various branches of the industry as the car industry. Among the almost 100 applications, the following sensors are currently in progress: portable digital display of the terrestrial magnetic field, brain tumor sensor, secondary current sensor for the induction motor control, car passing measurement and recording disk, finger-tip blood vessel pulsation, mechanoencephalogram sensor, etc[8, 159-161, 243].

On the other hand magnetic domain wall propagation becomes a hot topic of research because of possibility of applications in magnetic devices, such as Magnetic Random Access Memory, Integrated Circuits, Hard Disks, Domain-wall Logics, etc[134, 141, 244, 245]. High velocity of domain wall propagation observed in glass-coated microwires can be attractive to transmit the information along the magnetic microwire, like it was observed recently in wires of submicrometer diameter [12, 134, 141, 244]. Clear advantage of glass-coated microwires is that the domain wall velocity is few times higher, achieving few km/s [6, 14-16].

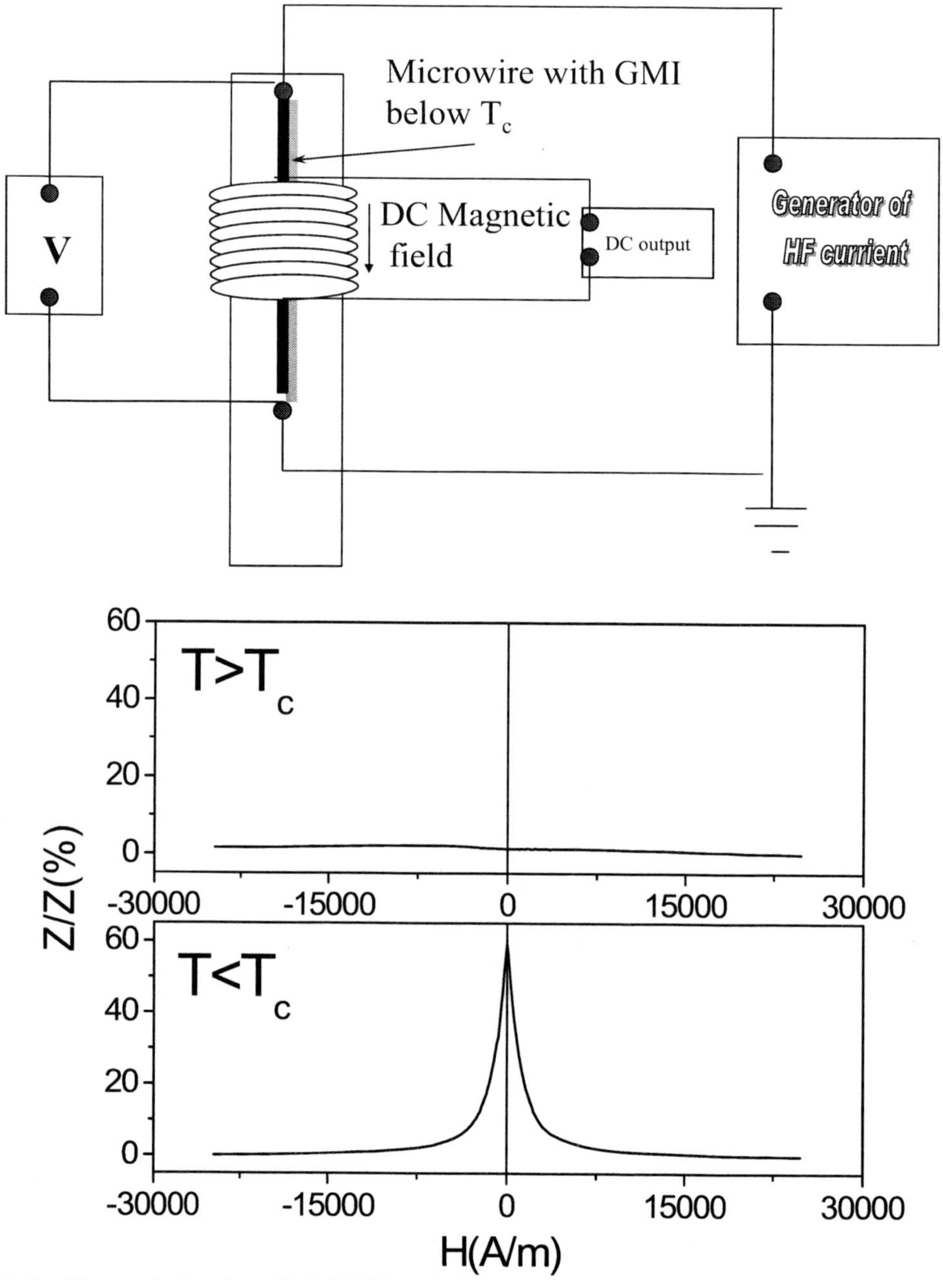

Reprinted with permission from Ref. [75] V. Zhukova et al., "Studies of magnetic properties of thin microwires with low Curie temperature", J. Magn. Magn. Mater, 300 16-23, 2006, Copyright Elsevier (2006) (Figure 9).

Figure 115. Schematic picture showing the working of the temperature sensor based on GMI effect in microwires with low T_C.

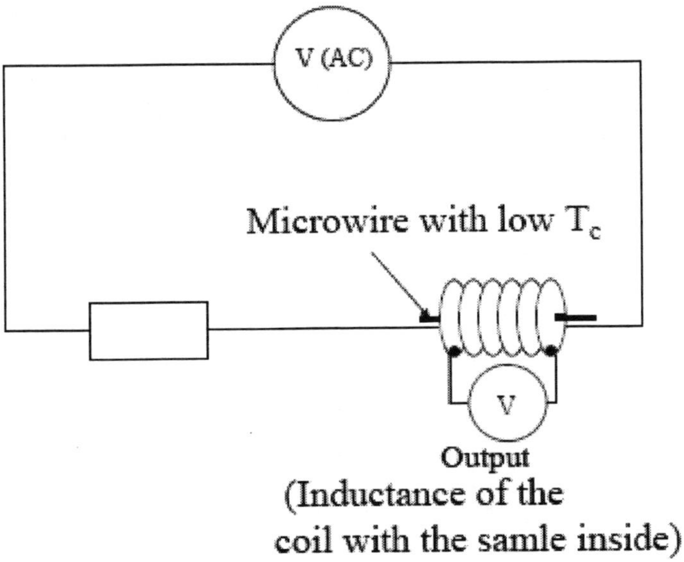

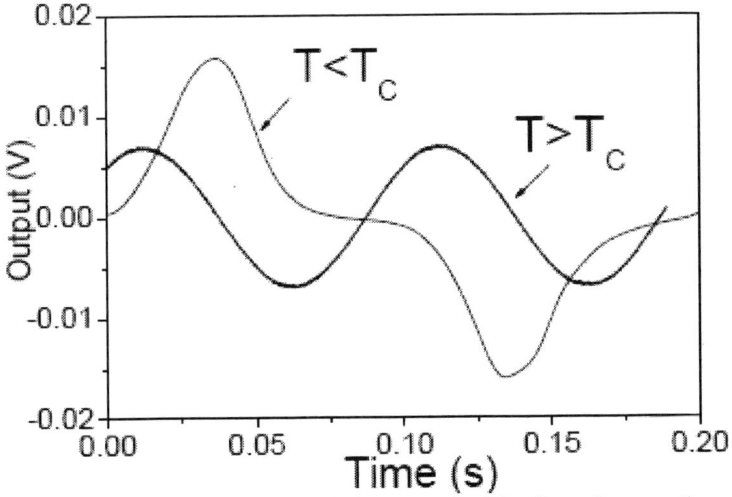

Reprinted with permission from Ref. [75] V. Zhukova et al., "Studies of magnetic properties of thin microwires with low Curie temperature", J. Magn. Magn. Mater, 300 16-23, 2006, Copyright Elsevier (2006) (Figure 10).

Figure 116. Schematic picture showing the working of the temperature sensor based on change of the inductance in microwires with low T_C.

There are many prototypes of magnetic sensors based on magneto-elastic properties, temperature dependence of magnetic properties, magnetic softness, and magnetic surveillance.

Few of such applications have been mentioned and explained in previous reviews [2, 119]. Among them magnetoelastic sensor of the level of liquid designed using the very sensible stress dependence of the hysteresis loop and particularly of the coercivity in nearly-zero magnetostriction CoMnSiB microwires [2, 119], magnetic surveillance system consisting on magnetic tags containing bistable microwires with plurality of coercivities, magnetoelastic sensor of air flux based on stress- sensitivity of the GMI effect, magneto-elastic pen etc [2,119].

Among recent applications, temperature dependence of the GMI effect and magnetic susceptibility in microwires with low Curie temperature can be mentioned [75,76].

Particularly, the functioning of the temperature sensor based on GMI effect is based on drastic drop of the GMI effect after achieving of the Curie temperature (see Figure 115). In this case the GMI effect of microwire disappears above Curie temperature, which is detected by the proposed circuitry. On the other hand the schematic picture showing the principle of the sensor based on the change of the inductance of the coil with the microwire inside is shown in Figure 116. In this case, the inductance of the coil with ferromagnetic wire inside drastically changes above Curie temperature of the microwire, allowing detection of the temperature.

On the other hand new types of stress- tuneable and temperature-tuneable composite materials based on thin ferromagnetic wires with the effective microwave permittivity depending on an external dc magnetic field, applied stress or temperature recently have been introduced [9, 194, 195]. Such composites consist of short pieces of conductive ferromagnetic wires embedded into a dielectric matrix (Figure 66). The short wire inclusions play a role of "the elementary scatterers", when the electromagnetic wave irradiates the composite and induces a longitudinal current distribution and electrical dipole moment in each inclusion. A number of applications have been proposed, including the stress-sensitive media for remote non-destructive health monitoring of different structures, temperature – dependent media and selective microwave coatings with field-dependent reflection/transmission coefficients [9, 194, 195]. The important advantage of such applications is that the soldering problems are avoided because of the wireless detection of the signals. It is worth mentioning that thin wires with stress sensitive magnetic anisotropy exhibiting stress sensitive GMI effect and SI effect are quite necessary for designing of such composites.

REFERENCES

[1] Jiles, D.C. *Acta Materialia* 2003, *51*, 5907-5939.

[2] Zhukov, A.; González, J.; Vázquez, M.; Larin, V. and Torcunov, A. *Nanocrystalline and Amorphous Magnetic Microwires,* Encyclopedia of Nanoscience and Nanotechnology, American Scientific Publishers: Valencia, USA, 2004; *6,* pp 365-387.

[3] Chiriac, H. and Ovari, T.A. *Progress in Material Science* 1996, *40,* 333-407.

[4] Zhukov, A.; Ipatov, M.; Zhukova, V.; García, C.; Gonzalez, J. and Blanco, J. M. *Phys. Stat. Sol. (A)*, 2008, *205, 6,* 1367-1372.

[5] Zhukov, A.; Gonzalez, J. and Zhukova, V. *J. Magn. Magn. Mater.* 2005, *294*, 165-173.

[6] Varga, R.; Zhukov, A.; Zhukova, V.; Blanco, J. M. and Gonzalez, J. *Phys. Rev. B* 2007, *76*, 132406

[7] Honkura, Y. in Advanced Magnetic Materials for Technolological Applications, 2008, pp 71-94, Editors: Arcady Zhukov and Julian Gonzalez, Transworld Research Network.

[8] Kanno, T.; Mohri, K.; Yagi, T.; Uchiyama, T.; Shen, L.P. *IEEE Trans.Magn.* 1997, *33, 5* 3358-3360

[9] Makhnovskiy, D. P.; Panina, L. V. and Sandacci, S. I. *Tuneable microwave composites based on ferromagnetic microwires*, Book chapter in Progress in Ferromagnetism Research, Editor: V. N. Murray, ISBN: 1-59454-335-6, Nova Science Publishers Inc., USA, 2005.

[10] Liu, L.; Kong, L. B.; Lin, G. Q.; Matitsine, S. and Deng, C. R. *IEEE Trans. Magn.* 2008 *44, 11,* 3119-3122.

[11] Beach, G. S. D.; Nistor, C.; Knutson, C.; Tsoi, M. and Erskine J. L. *Nat. Mater.* 2005 *4,* 741. Allwood, D. A.; Xiong, G.; Faulkner, C. C.; Atkinson, D.; Pctit, D. and Cowburn, R. P. *Science* 2005 *309*, 1688-1692.

[13] Zhukov, A. P.; Vázquez, M.; Velázquez, J.; Chiriac H. and Larin, V. *J. Magn. Magn. Mater.* 1995, *151,* 132-138.

[14] Zhukov, A. *Appl. Phys. Let.* 2001, *78*, 3106-3108.

[15] Varga, R.; Zhukov, A.; Blanco, J. M.; Ipatov, M.; Zhukova V.; Gonzalez, J.; Vojtaník, P. *Phys. Rev. B,* 2006, *74*, 212405.

[16] Zhukova, V.; Blanco, J.M.; Ipatov, M.; Varga, R.; Gonzalez, J.; Zhukov, A. *Physica B,* 2008, *403*, 282–285; Zhukova, V.; Blanco, J.M.; Ipatov, M.; Gonzalez, J. and Zhukov, A. "Domain wall propagation in thin Fe-rich glass-coated amorphous wires" *Phys. Status Solidi A* 2009, 206, No 4, 679-682 and Zhukova, V.; Blanco, J. M.; Ipatov, M.; Gonzalez, J. and Zhukov, A. "Domain Wall Propagation in Thin Fe-Rich Glass-Coated

Amorphous Wires" CP1003, *Magn. Mat., ICMM-2007*, p.301. Copyright © 2008 American Institute of Physics.

[17] Zhukova, V.; Cobeño, A.F.; Zhukov, A.; Blanco, J.M.; Puerta, S.; Gonzalez, J. and Vázquez, M. *Non-crystalline solids,* 2001, *287,* 31-36.

[18] Arcas, J.; Gómez-Polo, C.; Zhukov, A.; Vázquez, M.; Larin, V. and Hernando, A. *Nanostructured Materials*, 1996, *7*, No 8, 823-834.

[19] Miroshnichenko, I.S. and Salli, I.V. *Ind. Lab.,* 1959, *25,* 1463 in English*, Zav. Lab.* 1959, *25*, 1398.

[20] Duwez, P.; Williams, R.J. and Klement, K. *J. Appl. Phys.,* 1966, *31,* 1136.

[21] Duwez, P. in *Progress in Solid State Chemistry of Alloy Phases*, Ed. H. Reiss, Pergamon Press, Oxford, 1966, Vol.3, pp.377.

[22] Jones, H. *Rep. Progress Phys*., 1973, *36* 1425.

[23] Luborsky, F.E. *Amorphous Metallic Alloys* in Amorphous Metallic Alloys, Ed. By F. Luborsky, Butterworth and CoPublishers Ltd, 1983

[24] Gubanov, A.I. *Fizika,* 1960, *2*, 502.

[25] Mohri, K. and Tekeuchi, S. *IEEE Trans. Magn.* 1981, *17*, 3379-3381.

[26] Mohri, K. and Tekeuchi, S. *J. Appl.Phys.* 1982, *53*, 8386-8388.

[27] Davies, H. A. *Metallic glass formation* in Amorphous Metallic Alloys, Ed. By F. Luborsky, Butterworth and CoPublishers Ltd, 1983.

[28] Handrich, K. *Phys. Stat. Sol.,* 1969, *32,* K55

[29] Alben, R.; Becker, J.J. and Chi, M.C. *J. Appl. Phys.* 1978, *49,* 1653.

[30] Yoshizawa, Y.; Oguma, S. and Yamauchi, K. *J. Appl. Phys.,* 1988, *64,* 6044.

[31] Herzer, G. *IEEE Trans. Magn.* 1990, *26* 1397.

[32] Herzer, G. *Anales de Física B,* 1990, *86*, 157.

[33] Hernando, A. and Vázquez, M. in *Rapidly Solidified Alloys*, Ed. H.H. Liebermann, Marcel Dekker, New York, 1993, 553.

[34] Hernando, A.; Vázquez, M.; Kulik, T. and Prados, C. *Phys. Rev.B*, 1995, *51*, 3581.

[35] Hernando, A.; Marin, P.; Vázquez, M.; Barandiarán, J.M. and Herzer, G. *Phys. Rev.B,* 1998, *58,* 366.

[36] González, J.M.; De Julián, C.; Cebollada, F.; Giri, A.K. and González, J. *J. Appl. Phys.* 1997, *81,* 4658.

[37] Herzer, G. *Scr. Metall. Mater.* 1995, *33*, 1713.

[38] Varga, L.K.; Novák, L. and Mazaleyrat, F. *J. Magn. Magn. Mat.* 2000, *210*, L25.

[39] Morí, K. *IEEE Trans. Magn*. 1984, *20*, 942.

[40] Marín, P.; Vázquez, M.; Pascual, L.; Negri, D.; Leccabue, F.; Watts, B.E.; Davies, H.A. and Hernando, A. *Mater. Sci. Forum,* 1997, *235*, 743.

[41] Lim, S.H.; Pi, W.K.; Noh, T.H.; Kim, H.J. and Kang, I.K. *J. Appl. Phys.,* 1993, *73*, 6591.

[42] Tate, B. J.; Parmar, B. S.; Todd, I.; Davies, H. A.; Gibbs, M. R. J. and Major, R.V. *J. Appl. Phys.* 1998, *83,* 6335.

[43] Miguel, C.; Zhukov, A.; del Val, J.J.; Ramírez de Arellano, A.; Gonzalez, J. *J. Appl. Phys*., 2005, *97*, *3*, 034911.

[44] Frost, M.; Todd, I.; Davies, H.A.; Gibbs, M.R.J. and Major, R.V. *J. Magn. Magn. Mat.,* 1999, *203,* 85.

[45] Miguel, C.; Zhukov, A.P. and González, *J. Phys. Stat. Sol (a),* 2002, *194*, 291-303.

[46] Aranda, G.R.; Miguel, C.; García-Tello P. and González, J. *J. Appl. Phys.*, 2001, *89*, 6422.

[47] Panina, L.V.; Mizutani, M.; Mohri, K.; Humphrey, F.B. and Ogasawara, I. *IEEE Trans. Magn.* 1991, *27,* No 6, 5331-5333.

[48] Humphrey, F.B.; Mohri, K.; Yamasaki, J.; Kawamura, H.; Malmhäll, R.; Ogasawara, I. in *Magnetic Properties of Amorphous Metals*, Eds. A. Hernando, V. Madurga, M.C. Sánchez-Trujillo and M. Vázquez, Elsevier Science Publ., Amsterdam, 1987a, 110.

[49] Mohri K., Humphrey, F.B.; Kawashima, K.; Kimura, K. and Muzutani, M. *IEEE Trans. Magn.,* 1990, *Mag-26,* 1789.

[50] Vázquez, M.; Knobel, M.; Sánchez, M.L.; Valenzuela, R. and Zhukov, A. *Sensors and Actuators A,* 1997, *59,* 20.

[51] Panina, L.V. and Mohri, K. *Appl. Phys. Lett.,* 1994, *65*, 1189.

[52] Beach, R.S. and Berkowicz, A.E. *Appl. Phys. Lett.,* 1994, *64*, 3652.

[53] Vázquez, M.; Zhukov, A. *J. Magn. Magn. Mat.,* 1996, *160,* 223.

[54] Zhukov, A.; Vázquez, M.; Velázquez, J.; Chiriac, H. and Larin, V. *J. Magn. Magn. Mat.,* 1995, *151*, 132.

[55] Zhukov, A.; González, J.; Blanco, J.M.; Vázquez, M. and Larin, V. *J. Mat. Res.,* 2000, *15*, 210.

[56] Zhukov, A.; González, J.; Blanco, J.M.; Prieto, M.J.; Pina, E. and Vázquez, M. *J. Appl. Phys*., 2000, *87,* 1402.

[57] Zhukova, V.; Ipatov, M.; Gonzalez, J.; Blanco, J. M. and Zhukov, A. *J. Appl. Phys.* 2008, *103*, 07E714 -1-3 and Ipatov, M.; Zhukova, V.; Blanco, J. M.; Gonzalez, J. and Zhukov, A. *Phys. Stat. Sol. (A)* 2008, *205, No. 8,* 1779–1782.

[58] Zhukov, A.; Sinnecker, E.; Paramo, D.; Guerrero, F.; Larin, V.; González, J. and Vázquez, M. *J. Appl. Phys.,* 1999, *85*, 4482.

[59] Zhukov, A.; Martinez, J. L.; Zhukova, V.; Palomares, J.; Gonzalez, J.; del Val, J. J. and Vázquez, M. *IEEE Trans Magn.* 2004, *40, 4,* 2254-2256.

[60] Larin, V. S.; Torcunov, A. V.; Zhukov, A.; González, J.; Vazquez, M.; Panina, L. *J. Magn. Magn. Mater.* 2002, *249/1-2*, 39-45.

[61] Taylor G.F. *Phys. Rev.* 1924, *23*, 655-660.

[62] Taylor, G.F. *Process and apparatus for making Filaments*, Patented Feb. 24, 1931, United States Patent Office, 1, 793, 529.

[63] Ulitovsky, A.V. in *Micro-technology in design of electric devices*, Leningrad, 1951, No.7, p.6.

[64] Ulitovski, A.V. and Avernin, N.M. *Method of fabrication of metallic microwire* Patent No161325 (USSR), 19.03.64. Bulletin No7, p.14.

[65] Ulitovsky, A.V.; Maianski, I.M.; Avramenco, A.I. *Method of continuous casting of glass coated microwire* Patent No128427 (USSR), 15.05.60. Bulletin. No10, p.14.

[66] Badinter, E.Ya.; Berman, N.R.; Drabenko, I.F.; Zaborovsky, V.I.; Zelikovsky, Z.I. and Cheban, V.G. *Cat micwories and its properties*, Shtinica, 1973, Kishinev.

[67] Balyuk, Z.V. *Modelling, study of mechanism of high velocity crystallization on crystalline and glass-like substrates and peculiarities of phase and structure formation of cast microwire.* PhD, Dnepropetrovsk, 1997.

[68] Zelikovski, Z.I.; Zotov, S.K.; Nesterovski, I.A. *"On conditions of simultaneous extraction of metallic nucleus and glass capillary during the microwire fabrication"* in

"Microwire and electrical resistance devices" No. 8, Kishinev, "Cartia Moldavenske", 1971, pp. 3-8
[69] Zotov, S.K.; Kabisov, K.S.; Silkes, I.G. *Formation of parameters of cast microwires,* "Microwire in the resistance devices", Kishinev, *"Cartia Moldavenske",* 1974, pp. 3-17.
[70] Larin, V. S.; Torcunov, A. V.; Zhukov, A.; González, J.; Vázquez, M. and Panina, L. *J. Magn. Magn. Mater.* 2002, *249/1-2,* 39 ()
[71] Zhukov, A.; Luna, C.; Martinez, J.L.; Zhukova, V. and Vázquez, M. *J. Magn. Magn. Mat.* 2004, *272-276,* e1389-e1391.
[72] Zhukov, A.; Martín y Marero, D.; Batallan, F.; del Val, J. J.; Zhukova, V.; Martinez, J. L.; Luna,C.; Gonzalez, J.; Kaloshkin, S. and Vázquez, M. *Phys. Stat. Sol. (C)* 2004, *1, 12*, 3717-3721.
[73] Zhukov, A.; García, C.; Zhukova, V.; Larin, V.; González, J.; del Val, J. J.; Knobel, M.; Blanco, J. M. *Journal of Non-Crystalline Solids* 2007, *353*, 922–924.
[74] García, C.; Zhukov, A.; González, J.; del Val, J. J.; Blanco, J.M. and Zhukova, V. *Alloys and Compounds,* (accepted 2008).
[75] Zhukova, V.; Ipatov, M.; Zhukov, A.; Varga, R.; Torcunov, A.; Gonzalez, J. and Blanco, J.M. *J. Magn. Magn. Mater.* 2006, *300*, 16-23.
[76] Zhukova, V.; Blanco, J.M.; Ipatov, M.; Zhukov, A.; Garcıa, C.; Gonzalez, J.; Varga, R.; Torcunov,A. *Sensors and Actuators B* 2007, *126*, 318–323.
[77] Nixdorf, J. *Draht-Welt,* 1967, *53*, 696.
[78] Goto, T.; Nagano, M. and Wehara, N. *Trans JIM,* 1977, *18,* 759.
[79] Goto, T. *Trans JIM,* 1978, *19*, 60.
[80] Goto, T. *Trans JIM,* 1980, *21*, 219.
[81] Cobeño, A.F.; Zhukov, A.; de Arellano-López, A.R.; Elías, F.; Blanco, J.M.; Larin, V. and González, J. *J. Mat. Res.* 1999, *14*, 3775
[82] Zhukova, V.; Cobeño, A.F.; Zhukov, A.; de Arellano Lopez, A. R.; López-Pombero, S.; Blanco, J.M.; Larin, V. and Gonzalez, J. *J. Magn. Magn. Mater.* 2002, *249*, 79.
[83] Zhukova, V.; Zhukov, A.; Larin, V.; Torcunov, A.; Gonzalez, J.; de Arellano Lopez, A. R.; Quispe-Cancapa, J.J. and Pinto-Gómez, A.R. *Materials Science Forum* 2005, *480-481*, 293-297.
[84] Kraus L., Schneider J.and Wiesner H., *Czech. J. Phys.* 1976, *B26,* 601, H. Wiesner and J. Schneider, Phys. Stat. Sol. (a) 1974, 26, 71 - 75.
[85] Antonov, A.S.; Borisov, V.T.; Borisov, O.V.; Prokoshin, A.F. and Usov, N.A. *J. Phys. D: Appl. Phys.* 2000, *33*, 1161-1168.
[86] Catalan, C.F.; Prida, V.M.; Alonso, J.; Vázquez, M.; Zhukov, A.; Hernando, B. and Velázquez, J. *Materials Science and Engineering A, Supplement* 1997, 438-441.
[87] Garcia Prieto, M.J.; Pina, E.; Zhukov, A.P.; Larin, V.; Marin, P.; Vázquez, M. and Hernando, A. *Sensors and Actuators A* 2000, *81/1-3,* 227-231.
[88] Olivera, J.; Provencio, M.; Prida, V.M.; Hernando, B.; Santos, J.D.; Pérez, M.J.; Gorria, P.; Saánchez, M.L.; Belzunce, F.J. *J. Magn. Magn. Mater.* 2005, *294*, e163–e166.
[89] Zhukov, A.; Vázquez, M.; Velázqez, J.; Hernando, A. and Larin, V. *J. Magn. Magn. Mat.* 1997, *170*, 323-330.
[90] Gudoshnikov, S. A.; Grebenshchikov, Yu. B.; Ljubimov, B. Ya.; Palvanov, P. S.; Usov, N. A.; Ipatov, M.; Zhukov, A. and Gonzalez, J. *Phys. Stat. Sol (A)* 2009, 206, No.4, 613-617.

[91] García, C.; Zhukov, A.; Gonzalez, J.; Zhukova, V. and Blanco, J. M. *Trans. Magn. Soc. Jpn.* 2005, *5 No 4*, 148-151.

[92] Zhukov, A.; Gonzalez, J.; Blanco, J.M.; Aragoneses, P. and Dominguez, L. *Sensors and Actuators A* 2000, *81/1-3,*129-133.

[93] Chiriac, H.; Óvari, T.A. and Pop, Gh. *J. Magn. Magn. Mat.* 1996, *157/158*, 227.

[94] Aragoneses, P.; Blanco, J.M.; Cobeño, A.F.; Dominguez, L.; Gonzalez, J.; Zhukov, A. and Larin, V. *J. Magn. Magn Mat.* 1999, *196-197,* 248-250.

[95] Aragoneses, P.; Blanco, J.M.; Dominguez, L.; González, J.; Zhukov, A. and Vázquez, M. *J. Phys. D: Applied Phys.* 1998, *31,* 3040-3045.

[96] Cobeño, A.F.; Aragoneses, P.; Blanco, J.M.; Dominguez, L.; Zhukov, A.; Larin, V. and González, J. *Mat. Science Forum*, 1999, *302-303,* 244-248.

[97] Chiriac, H.; Pop, Gh.; Baraiu, F.; Óvari, T.A. and Tomut, M. *J. Non-cryst.Solids* 1996, *205-207,* 687-689.

[98] Severino, A.M.; Gómez-Polo, C.; Marin, P. and Vázquez, M. *J. Magn. Magn. Mat* 1992, *103*, 117-125.

[99] Cobeño, A. F.; Blanco, J. M.; Zhukov, A.; González, J. *J. Magn. Magn. Mat.* 2002, *249/1-2,* 396-400.

[100]Barandiarán, J.M.; Hernando, A.; Madurga, V.; Nielsen, O.V.; Vázquez, M. and Vázquez - López, M. *Phys. Rev. B.* 1987, *35* 5066.

[101]Blanco, J. M.; Barbon, P.G.; González, J.; Gomez-Polo, C. and Vázquez, M. *J. Magn. Magn. Mater.* 1992, *104-107,* 137.

[102]Zhukova, V.; Usov, N.A.; Zhukov, A. and Gonzalez, J. *Phys. Rev B*, 2002, *65*, 134407-1-7.

[103]Zhukova, V.; Zhukov, A.; Blanco, J.M.; Usov, N. and Gonzalez, J. *J. Magn. Magn. Mater.* 2003, *258-259*, 189-191.

[104]Zhukova, V.; Zhukov, A.; Blanco, J.M.; Gonzalez, J.; Gómez –Polo, C. and Vázquez, M. *J. Appl. Phys.* 2003, *93*, 7208-7210.

[105]Chen, D.X.; Gomez-Polo, C. and Vazquez, M. *J.Magn.. Magn. Mat.,* 1993, *124,* 262.

[106]Zhukova, V.; Zhukov, A.; Blanco, J.M.; Gonzalez, J. and Ponomarev, B. K. *J. Magn. Magn. Mat.* 2002, *249/1-2,* 131-135.

[107]Heiden, C. and Rogalla, H. *J. Magn. Magn. Mater.* 1982, *26*, 275.

[108]Ponomarev, B.K.; Zhukov, A.P. *Sov. Phys. Solid state*, 1984, *26*, 2874.

[109]Zhukov, A. *Materials and Design* 1993, *14*, 299.

[110]Ponomarev, B.K.; Zhukov, A. P. *Sov. Phys. Solid state* 1985, *27*, 444.

[111]Gawronski, P.; Zhukov, A.; Zhukova, V.; Blanco, J. M.; González, J. and Kulakowski, K. *Appl. Phys. Lett.* 2006, *88,* 152507.

[112]Varga, R.; Zhukov, A.; Ipatov, M.; Blanco, J. M.; Gonzalez, J.; Zhukova, V.; Vojtaník, P. *Phys. Rev. B* 2006, *73*, 054408-1-5 and Vázquez, M.; Zhukov, A.; Pirota, K.R.; Varga, R.; Hernandez, M.; García, K.L.; Luna, C.; Provencio, M.; Navas, D. and Martínez, J.L. *J. Non-cryst. Solids* 2003, *329 (1-3)*, 123-130.

[113]Varga, R.; Zhukov, A.; Blanco, J. M.; Gonzalez, J.; Zhukova, V. and Vojtanik, P. *Physica B: Condensed Matter* 2006, *V372*, 230-233 and Varga, R.; García, K.L.; Zhukov A. and Vázquez M. and Vojtanek, P. *Appl. Phys. Lett.* 2003, *83 13*, 2620-2622.

[114]Varga, R.; García, K.L.; Vázquez, M.; Zhukov, A.; Vojtanik, P. *Phys. Rev. B* 2004, *70*, 024402-1 and Varga, R.; García, K.; Luna, C.; Zhukov, A.; Vojtanik, P. and Vázquez, M. *Distribution and temperature dependence of switching field in bistable magnetic*

amorphous microwires; in Recent Res. Devel. Non-Crystalline Solids; Publisher: Transworld Research Network, Trivandrum, In, 2003; Vol. 3, pp 85-95, ISBN: 81-7895-090-1.
[115]Reininger, T.; Kronmüller, H.; Gomez-Polo, C. and Vazquez, M. *J. Appl. Phys.* 1993, *73*, 5357.
[116]Varga, R.; Zhukov, A.; Blanco, J. M.; Gonzalez, J.; Zhukova, V. and Vojtanik, P. *J.Magn. Magn. Mater.* 2006, *300*, e305-e-310.
[117]Jiles, D. *Introduction to Magnetism and Magnetic Materials* (Chapmen and Hall, London, 1991)
[118]H. Kronmüller, *J. Magn. Magn. Mater.* 1981, *24,* 159.
[119]Zhukov, A. and Gonzalez, J. *Amorphous and Nanocrystalline Soft Magnetic Materials*, in Advanced Magnetic Materials, book 3 "Processing of advanced magnetic materials." Edited by Liu, Y.; Sellmyer, D.J. and Shindo, D.; Kluwer Academic Publishers, Norwell, MA, USA, 2004; Vol. 3, Capítulo 5, pp 115-181.
[120]O'Handley, R. C. *Phys. Rev. B* 1978, *18*, 930.
[121]Hernando, A.; Madurga, V.; Núñez de Villavicenco, C.; Vazquez, M. *Appl. Phys. Lett.* 1984, *45*, 802.
[122]O'Handley, R.C. *Sol. State Comm.* 1977, *22*, 485.
[123]Vazquez, M.; Hernando, A.; Nielsen, O.V. *J. Magn. Magn. Mater.* 1986, *61*, 390.
[124]Madurga, V.; Vazquez, M.; Hernando, A.; Nielsen, O.V. *Sol. State Comm.* 1984, *52*, 701.
[125]Sato Turtelli, R.; Holzer, D.; Grössinger, R.; Sassik, H.; Pillmayer, N. *J. Magn. Magn. Mater.* 2001, *226-230*, 1496.
[126]Guntzel, U.; Westerholt, K.; Methfessel, S. *J. Magn.. Magn. Mater.* 1983, *38*, 294.
[127]Kronmüller, H. *J. Magn. Magn. Mater.* 1984, *41*, 366.
[128]Kronmüller, H. *Phys. Stat. Sol. (B)* 1985, *127*, 531.
[129]Varga. R. and Vojtanik, P. *J. Magn. Magn. Mater.* 1999, *196-197,* 230.
[130]Yamasaki, J.; Takajo, M. and Humphrey, F.B. *IEEE Trans. Magn.* 1993, *29*, 2545.
[131]Parkin, S. S. "System and Method for Writing to a Magnetic Shift Register" U.S. Patent 6898132, 2005
[132]Chen, D.-X.; Dempsey, N. M.; Vázquez, M. and Hernando, A. *IEEE Trans. Magn.* 1995, *31*, 781.
[133]Riehemann, W. and Nembach, E. *J. Appl. Phys.* 1984, *55*, 1081.
[134]Parkin, S. S. P. U.S. Patent No. US 683 400 5, 2004.
[135]Kunza, A.; Reiff, S. C. *J. Appl. Phys.* 2008, *103*, 07D903.
[136]Zhukov, A. *J. Magn. Magn. Mater.* 2002, *242-245*, 216.
[137]Kabanov, Yu.; Zhukov, A.; Zhukova, V.; Gonzalez, J. *Appl. Phys. Let.* 2005, *87*, 142507.
[138]Sixtus, K.J. and Tonks, L. *Phys. Rev.* 1932, *42*, 419.
[139]Williams, H.J.; Shockley, W.; Kittel, C. *Phys. Rev.* 1950, *80*, 1090.
[140]Himeno, A.; Ono, T.; Nasu, S.; Okuno, T.; Mibu K. and Shinjo, T. *J. Magn. Magn. Mater.* 2004, *272-276*, 1577.
[141]D. Atkinson, D.A. Allwood, G. Xiong, M.D. Cooke, C.C. Faulkner and R.P. Cowburn, *Nature Mater*. 2, 85 (2003)
[142]Varga, R.; García, K.L.; Vázquez, M.; Vojtanik, P. *Phys. Rev. Lett.* 2005, *94,* 017201.
[143]Yang, S.; Erskine, J.L. *Phys. Rev. B* 2005, *72*, 064433.

[144]Kostyk, Y.; Varga, R.; Vazquez, M.; Vojtanik, P. *Physica B* 2008, *403*, 386.

[145]Forster, H.; Schrefl, T.; Scholz, W.; Suess, D.; Tsiantos, V. and Fidler, J. *J. Magn. Magn. Mater.* 2002, *249*, 181.

[146]McMichael, R.D. and Donahue, M.J. *IEEE Trans. Magn.* 1997, 33, 4167.

[147]Sethna, J.P.; Dahmen, K.A. and Perkovic, O. In *The Science of Hysteresis;* ed. Bertotti, G. and Mayergoyz, I.; Random Field Ising Models of Hysteresis; Publisher: Academic Press, 2006 p 181.

[148]Durin, G. and Zapperi, S. In *The Science of Hysteresis;* ed. Bertotti, G. and Mayergoyz, I.; Barkhaussen Effect; Publisher: Academic Press, 2006 p 181.

[149]Ipatov, M.; Usov, N.A.; Zhukov, A.; Gonzalez, J. *Physica B* 2008, *403*, 279–281.

[150]Ipatov, M.; Blanco, J.M.; Zhukova, V.; Zhukov, A. (to be published)

[151]Varga, R.; Richter, K.; Zhukov, A. and Larin, V. *IEEE Trans. Magn.* 2008, *44, 11,* 3925-3930.

[152]Chiriac, H.; Hristoforou, E.; Neagu, M.; Darie, I.; Moga, A.E. *J. Non-Cryst. Solids* 2001, *287*, 413-416.

[153]Chiriac, H.; Hristoforou, E.; Neagu, M.; Darie, I.; Hison, C. *Sensors and Actuators* 2000, *81*, 150–153.

[154]Demokritov, S.O.; Kirilyuk, A.; Kreines, N.M.; Kudinov, V.; Smirnov, V.B.; Chetkin, M.V. *J. Magn. Magn. Mater.* 1991, *102*, 339-353.

[155]Chetkin, M. V.; Kurbatova, Yu. N.; Shapaeva, T. B. and Borshchegovskii, O. A. *J. Exp. Theor. Phys.* 2006, 103, 159–164.

[156]Chetkin, M.V.; Kurbatova, Yu.N.; Shapaeva, T.B.; Borschegovsky, O.A. *Physics Letters A* 2005, *337*, 235–240.

[157]Schryer, N.L. and Walker, L.R. *J. Appl. Phys.* 1974, *45*, 5406-5421.

[158]Nakatani, Y.; Thiaville, A. and Miltat, J. *Nature Mater.* 2003, *2*, 521-523.

[159]Thiaville, A.; Garcia, J.M. and Miltat, J. *J. Magn. Magn. Mater.* 2002, *242-245*, 1061-1063.

[160]Zhukova, V.; Chizhik, A.; Zhukov, A.; Torcunov, A.; Larin, V. and Gonzalez, J. *IEEE Trans. Magn.* 2002, *38, 5, partI*, 3090-3092.

[161]Zhukov, A.; Zhukova, V.; Blanco, J.M. and Gonzalez, J. *J. Magn. Magn. Mater.* 2005, *294*, 182-192.

[162]Herzer, G.; Vazquez, M.; Knobel, M.; Zhukov, A.; Reininger, T.; Davies, H.A. and Grössinger, R. *J. Magn.. Magn.. Mater.* 2005, *294*, 252-266.

[163]Zhukov, A.; Zhukova, V.; Blanco, J.M. and Gonzalez, J. *The Physics of Metalls and Metallography* 2005, *99, Suppl.1*, s57-s61.

[164]Zhukova, V.; Kaloshkin, S.; Zhukov, A. and Gonzalez, J. *J. Magn. Magn. Mat.* 2002, *249 /1-2,* 108-112.

[165]Shen, L.P.; Uchiyama, T.; Mohri, K.; Kita, E.; Bushuda, K. *IEEE Trans.Magn.* 1997, *Mag-33,* 3355.

[166]Mohri, K.; Uchiyama, T.; Shen, L.P.; Cai, C.M.; Panina, L.V. *Sensors and Actuators A*, 2001, *91*, 85.

[167]Honkura, Y. *J. Magn. Magn.. Mater.* 2002, *249*, 375.

[168]Usov, N.A.; Antonov, A.S. and Lagar`kov, A.N. *J. Magn. Magn. Mat.* 1998, *185*, 159.

[169]Yoshinaga, Y.; Furukawa, S. and Mohri, K. *IEEE Trans.Magn.* 1999, *35,* 3613.

[170]Blanco, J. M.; Zhukov, A. and Gonzalez, J. *J. Magn. Magn. Mat.* 1999, *196-197,* 377-379.

[171]Blanco, J.M.; Zhukov, A. and Gonzalez, J. *J. Phys. D:Appl. Phys.* 1999, *32*, 3140-3145.

[172]González, J.; Chen, A.P.; Blanco, J.M. and Zhukov, A. *Phys. Stat. Sol.* 2002, *189*, 599-608.

[173]Knobel, M.; Vázquez, M. and Kraus, L. *Giant Magnetoimpedance*, in Handbook of Magnetic Materials, vol 15, Ed. K.H.K. Buschow, Elsevier, 2003, p.497.

[174]Makhnovskiy, D. P.; Panina, L.V. and Mapps, D. J. *Phys.Rev.B,* 2001, *63*, 1444241.

[175]Zhukova, V. A.; Chizhik, A. B.; Gonzalez, J.; Makhnovskiy, D. P.; Panina, L. V.; Mapps, D. J. and Zhukov, A. P. *J. Magn.. Magn.. Mat.* 2002, *249*, 318.

[176]Narita, K.; Yamasaki, J. and Fukunaga, H. *IEEE Trans. Magn.* 1980, *Mag-16,* 435.

[177]Siemko, A. and Lachowicz, H. *IEEE Trans. Magn.* 1987, *Mag-23*, 2563.

[178]Hernando, A.; Vázquez, M.; Madurga, V.; Kronmüller, H. *J. Magn. Magn. Mat.* 1983, *37*, 161.

[179]Vázquez, M.; Fernengel, W. and Kronmüller, H. *Phys. Stat. Sol.* 1983, *80*, 195.

[180]Zhukov, A.; Zhukova, V.; Blanco, J. M.; Cobeño, A. F.; Vazquez, M. and Gonzalez, J. *J. Magn.. Magn. Mater.* 2003, *258-259,* 151-157.

[181]Cobeño, A.F.; Zhukov, A.; de Arellano-Lopez, A.R.; Elías, F.; Blanco, J.M.; Larin, V. and González, J. *J. Mat. Res.* 1999, *14*, 3775-3783.

[182]Gonzalez, J.; Zhukov, A. P.; Blanco, J. M.; Cobeño, A. F.; Vázquez, M. and Kulakowski, K. *J. Appl. Phys.* 2000, *87 No 9*, 5950-5952.

[183]Zhukova, V.; Blanco, J.M.; Zhukov, A.; Gonzalez, J.; Torcunov, A. and Larin, V. *J. Magn. Magn. Mater.* 2003, *254-255*, 94-96.

[184]Zhukova, V.; Blanco, J.M.; Zhukov, A.; Gonzalez, J. *J. Phys. D: Appl. Phys.* 2001, *34*, L113-L116.

[185]Hilzinger, H.R.; Hillmann, H. and Mager, A. *Phys .Stat. Solidi (A)*, 1979, *55*, 763.

[186]Zhukova, V.; Ipatov, M.; García, C.; Gonzalez, J.; Blanco, J. M. and Zhukov, A. *Open Materials Science Reviews,* 2007, *1,* 1-12.

[187]Zhukov, A. *Adv. Func. Mat.* 2006, *16, Issue 5*, 675-680.

[188]Zhukov, A.; Zhukova, V.; Larin, V.; Blanco, J.M. and Gonzalez, J. *Physica B* 2006, *384*, 1-4.

[189]Larin, V. and Zhukov, A. *Phys. Stat. Sol. (C)* 2009, *6, No 4*, 958-961.

[190]O'Handley, R. C.; Hasegawa, R.; Ray, R. and Chou, C.-P. *Appl. Phys. Lett.* 1976, *29*, 330.

[191]Sandacci, S. I.; Makhnovskiy, D. P.; Panina, L. V.; Mohri, K. and Honkura, Y. *IEEE Trans Magn.* 2004, *35*, 3505.

[192]Antonov, A.S.; Iakubov, I.T.; Lagarkov, A.N. *J. Magn. Magn. Mat.* 1998, *187*, 252.

[193]Zhukova, V.; Blanco, J.M.; Ipatov, M.; Zhukov, A.; Garcıa, C.; Gonzalez, J.; Varga, R.; Torcunov, A. *Sensors and Actuators B* 2007, *126*, 318–323.

[194]Zhukov, A.; Zhukova, V.; Gonzalez, J.; Panina, L. and Blanco, J.M. *Advances in Science and Technology* 2008, *54,* 180-186.

[195]Makhnovskiy, D.; Zhukov, A.; Zhukova, V.; Gonzalez, J. *Advances in Science and Technology* 2008, *54,* 201-210.

[196]Chizhik, A.; Zhukov, A.; Blanco, J.M. and Gonzalez, J. *Non-crystalline solids* 2001, *287*, 374-379.

[197]Chizhik, A.; Zhukov, A.; Blanco, J.M.; Szymczak, R. and Gonzalez, J. *J. Magn. Magn. Mat.* 2002, *249/1-2,* 99-103.

[198]Sampaio, L.C.; Sinnecker, E.H.C.P.; Cernicchiaro, G.R.C.; Knobel, M.; Vázquez, M. and Velázquez, J. *Phys.Rev.B* 2006, *1,* 8976.

[199]Sinnecker, J.P.; de Araujo, A.E.P.; Piccin, R.; Knobel, M. and Vázquez, M. *J. Magn. Magn. Mat.* 2005, *295*, 121.

[200]Zhukov, A.; Vázquez, M.; Velázquez, J.; Garcia, C.; Valenzuela, R. and Ponomarev, B. *J. Mat. Sci. Eng.* 1997, *A226-228*, 753-756.

[201]García, C.; Zhukova, V.; Gonzalez, J.; Blanco, J.M.; Zhukov, A. *Physica B* 2008, *403*, 286–288.

[202]García, C.; Zhukova, V.; Zhukov, A.; Usov, N.; Ipatov, M.; Gonzalez, J. and Blanco, J.M. *Sensor Letters,* 2007, *5, Number 1*, 10-12.

[203]Ubizskii, S.B.; Pavlyk, L.P. *Sensors and Actuators A* 2008, *141*, 440-446.

[204]Moron, C.; Aroca, C.; Sanchez, M.C.; Garcıa, A.; Lopez, E. and Sanchez, P. *IEEE. Trans. Magn.* 1995, *31*, 906-909.

[205]Samsonova, V.; Ilin, M.; Ipatov, M.; Zhukova, V.; Perov, N. and Zhukov, A. (to be published).

[206]Pirota, K.; Hernandez-Velez, M.; Navas, D.; Zhukov, A.; Vázquez, M. *Adv. Funct. Mat.* 2004, *14*, 266-268.

[207]Pirota, K.; Provencio, M.; García, K.; Hernandez-Velez M. and Vázquez M. *J. Magn. Magn. Mat.* 2005, *290-291* , 68.

[208]Vázquez, M.; Pirota, K.; Torrejon, J.; Badini, G.; Torcunov, A. *J. Magn. Magn. Mat.* 2006, *304*, 197-202 and Torrejón, J.; Badini, G.; Pirota, K.; Vázquez, M. *Acta Materialia* 2007, *55,* 4271–4276.

[209]Zhukova, V.; Cobeño, A.F.; Zhukov, A.; Blanco, J.M.; Larin, V. and González, J. *Nanostructured Materials,* 1999, *11*, 1319.

[210]González, J.; Zhukov, A.; Zhukova, V.; Cobeño, A.F.; Blanco, J.M.; de Arellano-López, A.R.; López-Pombero, S.; Martínez-Fernández, J.; Larin, V. and Torcunov, A. *IEEE Trans. Magn.* 2000, *Mag-36*, 3015,

[211]Garcia, C.; Zhukov, A.; Ipatov, M.; Zhukova, V.; del Val, J.J.; Domínguez, L.; Blanco, J.M.; Larin, V. and González, J. *Journal of Optoelectronics and Advanced Materials*, 2006, *8, No 5*, 1667-1671.

[212]García, C.; Zhukov, A.; Gonzalez, J.; Zhukova, V.;. Varga, R; del Val, J.J.; Larin, V.; Chizhik, A. and Blanco, J.M.; *J. Appl. Phys.* 2006, *99*, 08F116.

[213]García, C.; Zhukov, A.; Gonzalez, J.; Zhukova, V.; Varga, R.; del Val, J.J.; Larin, V. and Blanco, J.M. *J. Alloys and compounds*, 2006, *423*, 116-119.

[214]Dudek, Ch.; Adenot-Engelvin, A-L. and Acher, O. *IEEE Trans. Magn.* 2006, *42, No 10*, 2787.

[215]Baibich, M.N.; Broto, J.M.; Fert, A.; Nguyen Van Dau, F.; Petroff, F.; Etienne, P.G.; Creucet, P.; Friederich, A. and Chazelas, J. *Phys. Rev. Lett.* 1988, *61*, 2472.

[216]Berkowitz, A.E.; Mitchell, J.R.; Carey, M.J.; Young, A.P.; Zhang, S.; Spada, F.E.; Parker, F.T.; Hutten, A. and Thomas, G. *Phys. Rev. Lett.* 1992, *68,* 3745.

[217]Chien, C.L.; Xiao, J.Q. and Jiang, J.S. *J. Appl. Phys.* 1993, *73*, 5309.

[218]Yu, R.H.; Zhang, X.X.; Tejada, J.; Knobel, M.; Tiberto, P. and Allia, P. *J. Appl. Phys.* 1995, *78*, 392.

[219]Baibich, M.N.; Martínez, G.; Miranda , M.G.M.; da Rosa, A.T.; González, J.; Zhukov, A. *J. Magn. Magn. Mater.* 2008, *320,* e29-e31.

[220]Bozorth, R. M. *Ferromagnetism;* Van Nostrand, New York, 1951; p 402.

[221]Joffe, I. and Gaunt, P. *J. Phys. D: Appl. Phys.* 1971, *4*, 586.

[222]Zhukov, A.; García, C.; Del Val, J. J.; Gonzalez, J.; Knobel, M.; Serantes, D.; Baldomir, D. and Zhukova, V. *J. Phys. C* 2009, *21*, 035301.

[223]Gonzalez, J.; Zhukova, V.; Zhukov, A. P.; Del Val, J. J.; Blanco, J.M.; Pina, E. and Vázquez, M. *J. Magn. Magn. Mater.* 2000, *221*, 196-206.

[224]Peleg, N.; Shtrikman, S.; Gorodetsky, G.; Felner, I. *J. Magn. Magn. Mater.* 1999, *191*, 349.

[225]Denardin, J. C.; Brandl, A. L.; Knobel, M.; Panissod, P.; Pakhomov, A. B.; Liu, H. and Zhang, X. X. *Phys. Rev. B* 2002, *65*, 064422.

[226]Nunes, W. C.; De Biasi, E.; Meneses, C. T.; Knobel, M.; Winnischofer, H.; Rocha, T. C. R. and Zanchet, D. *Appl. Phys. Lett.* 2008, *92*, 183113.

[227]Del Val, J.J.; Gonzalez, J.; Zhukov, A. *Physica B* 2001, *299*, 242–250.

[228]Allia, P.; Knobel, M.; Tiberto, P.; Vinai, F. *Phys. Rev. B* 1995, *52*, 15398.

[229]Paulus, P.M.; Luis, F.; Kroll, M.; Schmid, G.; De Jongh, L.J. *J. Magn. Magn. Mater.* 2001, *224*, 180.

[230]Berkowitz, A. E.; Mitchell, J. R.; Carey, M. J.; Young, A. P.; Zhang, S.; Spada, F. E.; Parker, F. T.; Hutten, A. and Thomas, G. *Phys. Rev. Lett.* 1992, *68*, 3745.

[231]Berkowitz, A. E.; Mitchell, J. R.; Carey, M. J.; Young, A. P.; Rao, D.; Starr, A.; Zhang, S.;. Spada, F. E; Parker, F. T.; Hutten, A. and Thomas, G. *J. Appl. Phys.* 1993, *73*, 5320.

[232]Jin, X. F. *J. Phys. D: Appl. Phys.* 2002, *35*, 2332.

[233]Ilyn, M. I.; Zhukova, V.; Santos, J. D.; Sánchez, M. L.; Prida, V. M.; Hernando, B.; Larin, V.; González, J.; Tishin, A. M. and Zhukov, A. *Phys. Stat. Sol. (A)* 2008, *205, No 6,* 1378-1381.

[234]Pecharsky, V. K.; Gschneidner, K. A. *Int. Journ. of Refrigeration* 2006, *29*, 1239.

[235]Zhukov, A. P.; Ponomarev, B.K. *Fizika Tverdogo Tela* 1989, *31, 7,* 26.

[236]Jen, S.U.; Weng, C.J. *J. Appl. Phys.* 1988, *64,* 4627.

[237]Hozelitz, K. *J. Magn. Magn. Mater.* 1980, *20*, 201.

[238]Zhukova, V.; Cobeño, A.F.; Pina, E. Zhukov, A.; Blanco, J.M.; Dominguez, L.; Larin, V. and Gonzalez, J. *J. Magn. Magn. Mater.* 2000, *215-216,* 322-324.

[239]Aragoneses, P.; Blanco, J. M.; Kulakowski, K.; Dominguez, L.; Zhukov, A. and Gonzalez, J. *J. Phys.D:Applied Phys.* 1998, *31,* 494-497.

[240]Krupinska, G.; Zhukov, A.; Gonzalez, J. and Kulakowski, K. *Computational Material Science* 2006, *36*, 268-271.

[241]Zhukov, A.; Ipatov, M.; Gonzalez, J.; Blanco, J.M. and Zhukova, V. *J. Magn. Magn Mat.* 2008, *321*, 822-825.

[242]Zhukov, A.; Ipatov, M.; Blanco, J.M.; Gonzalez, J. and Zhukova, V. *Phys. Stat. Sol. (A)* 2009, 206, *4,* 674-678 (2009).

[243]Mohri, K.; Honkura, Y. *Sensor Lett.* 2007, *5, No2*, 267-270.

[244]Hayashi, M.; Thomas, L.; Rettner, C.; Moriya, R.; Jiang, X. and Parkin, S.S.P. *Phys. Rew. Lett.* 2006, *97*, 207205.

INDEX

A

B

C

D

E

S

T

U

V

W

X

Y